The
Geosystem

Geography
Oceanography Geodesy Climatology
Cartography

**THE GEOSYSTEM:
DYNAMIC INTEGRATION
OF LAND, SEA AND AIR**
George R. Rumney
University of Connecticut

History

**HISTORICAL GEOLOGY OF
NORTH AMERICA**
Morris Petersen
Keith Rigby
Lehi F. Hintze
Brigham Young University

**STRATIGRAPHY AND
GEOLOGIC TIME**
John W. Harbaugh
Stanford University

Physics

**GEOPHYSICS, GEOLOGIC
STRUCTURES AND TECTONICS**
John S. Sumner
University of Arizona

Chemistry

EARTH MATERIALS
Henry Wenden
Ohio State University

Engineering and Mining

APPLIED EARTH SCIENCE
Daniel S. Turner
Eastern Michigan University

BROWN
FOUNDATIONS
OF
EARTH
SCIENCE
SERIES

**Quaternary Studies and
Archeology**

**PLEISTOCENE GLACIATION
AND THE COMING OF MAN**
W. N. Melhorn
Purdue University

**Physical Geography and
Hydrology**

**LANDFORMS AND
LANDSCAPES**
Sherwood D. Tuttle
University of Iowa

Biology

**FOSSILS, PALEONTOLOGY
AND EVOLUTION**
David L. Clark
University of Wisconsin

Astronomy

**ASTRONOMY AND THE
ORIGIN OF THE EARTH**
Theodore G. Mehlin
Williams College

The Geosystem

Dynamic Integration
of Land, Sea and Air

BROWN FOUNDATIONS OF EARTH SCIENCE SERIES

George R. Rumney
University of Connecticut

WM. C. BROWN COMPANY PUBLISHERS
Dubuque, Iowa

FOUNDATIONS OF EARTH SCIENCE SERIES

Consulting Editor

SHERWOOD D. TUTTLE
University of Iowa

Printed in the United States of America.

Preface

In this book the earth is presented as a *geosystem.* Land, sea and air are dynamically integrated in a single, planetary system through processes by which energy, matter and momentum are continually exchanged. Solar radiation is the initial energy and the general circulation of the atmosphere is the primary integrating agent. The interaction processes are disciplined by the diurnal and annual rhythms arising from the motions of the earth. How those processes operate in obedience to the dictates of the earth's motions, as well as their relative effectiveness, provide most of the subject matter of this book. Thus its raw materials are commonly treated in the more specialized studies of climatology, geography, geology and oceanography. The book is intended to encourage the reader's interest in the general subject of earth science and at the same time to offer a global context within which he may pursue more fruitfully the component studies of that inclusive field.

The rotating sphere of earth is examined more and more frequently from space. Hence it is clearly appropriate at this time to synthesize the efforts of all branches of earth science. While the term geosystem is an innovation, the concept of global unity is not new. Beginning in the pre-Christian era with the investigations of Ionian Greek scholars such as Anaximander, Eratosthenes, and Posidonius, efforts to understand the geographical integration of the entire earth were revived, after a long interval of neglect, by the work of Varenius, Newton and others in the 17th century. The first modern theory of dynamic planetary unity was expressed by James Hutton in his *Theory of the Earth*, published in 1795. The *Geosystem* offers a contemporary treatise on the dynamic unity of the earth.

The author wishes to thank sincerely Dr. John V. Byrne, Oregon State University and Dr. Sherwood D. Tuttle, University of Iowa, for their valuable comments on the manuscript. The editorial advice of Professor J. C. P. Gauvin is acknowledged with deepest gratitude. The tireless patience of Jane Seeber in typing the manuscript is most deeply appreciated.

George R. Rumney

Contents

Man is a creature of the earth. Second to knowledge of man himself is the necessity to understand the earth. The comprehensive study of the earth and its phenomena is termed earth science. The subject area intersects numerous academic disciplines. To gain a thorough scientific understanding of the earth, one must study astronomy, meteorology, oceanography, geology, and geophysics, plus aspects of geography and engineering. Additionally, comprehension of these topics requires a prior knowledge of such areas as mathematics, physics, chemistry, and biology.

The knowledge explosion that has occurred during the twentieth century has made such an inclusive approach impossible for the educated layman. Nevertheless, the need to understand our earth has become increasingly necessary.

The Foundations of Earth Science Series, designed for use at the introductory level, incorporates into the scientific study of the earth an examination of the earth's components and their distribution, how and why they exist as they do, and how they affect civilized man.

A study of the Geosystem illustrates the concept of interdependence between air in the atmosphere, water in the seas, and effects of air and sea on the land, with all being acted upon by solar energy. An understanding of a local change in aspects of the Geosystem helps to explain dynamic processes worldwide.

Introduction

TOPICS

Purpose
The geosystem
The solar system
Earth motions

Latitude and longitude
Geographical foundations
Landforms

Essential to the aims of earth science is an understanding of the dynamic unity among land, sea and air. The integration of these primary planetary materials in a single, dynamically-unified system is the central theme of this book. The earth is thus regarded as an integrated geosystem. The geosystem, a component of the solar system, functions chiefly in accordance with the dictates of the earth's spherical form, its motions and the radiant heat of the sun. An understanding of the geosystem is approached by combining methods and data from astronomy, climatology, geography, geology, geophysics, and oceanography. In this volume stress is laid upon geography, climatology and oceanography.

In geography the significance of location is paramount. Spatial relationships between land and sea profoundly affect the operations of atmospheric, oceanic and terrestrial processes. It is thus necessary to know the size, shape and distribution of land masses, oceans and seas, as well as details of terrestrial landscapes and ocean basins. The world's geography is therefore a set of spatial relationships among details of varying magnitude dynamically organized in a system of never-ending change.

Climatology, like meteorology, is a science of the atmosphere. Meteorology is mainly concerned with the explanation of atmospheric processes toward the goal of reliable weather prediction. Climatology deals principally with the recorded results of atmospheric processes over a long period of time. Thus one of its aims is to provide a synthesis of all the known types of weather experienced over long intervals for the entire earth, for major segments of the earth, or for individual localities. Another is to explain the nature and distribution of the world's climates. And its methods permit an explanatory description of individual atmospheric elements in terms of normal state and departure from normal.

Oceanography, the science of the sea, endeavors to describe and explain the sea's physical and chemical properties. It stresses both internal qualities and external relationships with the land, the atmosphere and the teeming organic life of the sea. It is concerned with both generative processes and the results of those processes.

The snow-capped peaks of the Alps and the balmy warmth of the sea between Cape Verde and the West Indies may at first glance have little apparent connection. But the snows that swirl around Mont Blanc and the Matterhorn are part of the atmospheric moisture that is derived originally by evaporation from the sea, especially the sunlit waters of the tropical seas. It is relinquished through condensation when turbulent air is sufficiently cooled to produce precipitation that eventually falls to earth as rain or snow. The meltwater from those Alpine glaciers and snowfields, along with the rains of the lower slopes, feed the tributaries of four major rivers in Europe: the Rhine flowing into the North Sea, the Rhone emptying into the Mediterranean, the Po into the Adriatic, and the Danube into the Black Sea. Thus that moisture, originating in the sea, is returned to the sea. Furthermore, particles from the earth's solid surfaces in suspension, and soluble compounds in solution, are borne along by the seaward flow of the rivers. Hence the land yields some of its substance for transport into the sea. The cycle is continuous and is evident in the drainage features of every continent. In this way sea, air and land are united in a continual dynamic process in which the atmosphere plays the leading role.

THE GEOSYSTEM

Viewed from the moon, details of the earth's varied surface are obscured, and in the sun's rays land, sea and atmosphere merge into a luminous, mottled, circular image (Figure 1.1). The image changes slowly with the earth's rotation and the drifting pattern of its ephemeral clouds. But it is this view from space that suggests the dynamic unity of land, sea and air in a single *geosystem*.

The systematic unity of the earth's diversified surface is accomplished principally by the motions of air and sea and by the work of flowing water on land. In this the atmosphere is the chief integrating agent through the ceaseless operations of its general circulation.

Energy, matter and momentum are continuously interchanged (Figure 1.2). The geosystem is thus a self-modifying set of sub-systems. Material substances are constant, only changing the ways in which they are

Figure 1.1. The earth viewed from Apollo 8 about 22,000 miles in space. Although most of South America is cloud-covered, the north and west coasts are sharply defined, and the Caribbean Sea is virtually cloud-free under the westerly flow of the northeast trades. (NASA photo).

combined and distributed. The processes of integration are also constant, only changing location and intensity as the geography of the planet changes. The earth's geography is constantly changing and its existing state is simply the most recent of a very long series of developmental stages. Thus the geosystem of the present is only the latest in a succession of geosystems, destined to give way through self-modification to future geosystems.

To understand the processes by which the geosystem operates, a grasp is required of the earth's relationships within the solar system and the foundational geography of its lands and seas.

THE SOLAR SYSTEM

This is the aggregate of planets, satellites, asteroids, comets and meteors that revolve methodically around the sun. The heliocentric system is the parent mechanism of which the earth's geosystem is a part. In turn, the sun's family is only a very small cluster in the unimaginably vast, revolving spiral of celestial bodies known as the Milky Way galaxy.

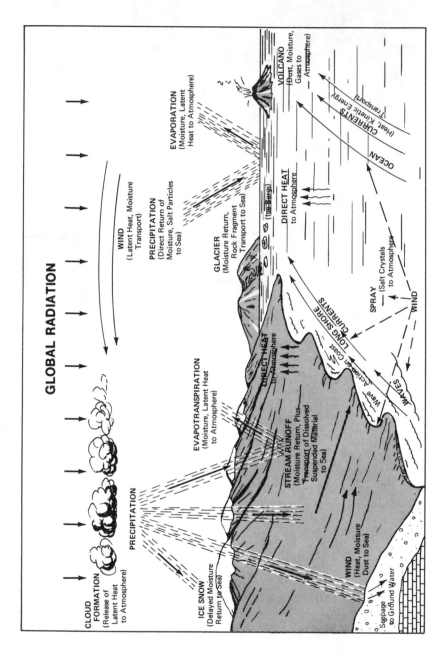

Figure 1.2. The processes of the geosystem by which energy, matter and momentum are continuously exchanged among land, sea and air.

4

The spinning sphere of the earth, about 8,000 miles in diameter hence about 25,000 miles in circumference, possesses a shape technically termed an *oblate spheroid* (Figure 1.3). This arises from a slight flattening at the poles resulting in a polar diameter (7,900 miles) that is about 27 miles shorter than its equatorial diameter of 7,926.7 miles. Consequently the planet's polar circumference is also smaller, measuring 24,619 miles in contrast to its equatorial circumference of 24,903 miles. In view of these very slight differences it has long been the practice to speak of the earth as a sphere.

The spherical earth completes a single rotation on its axis once every 24 hours and about 4 minutes. The speed at which points along the equator move with rotation is about 1,038 miles per hour. With increasing distance from the equator, the rotational speed of the earth's surface diminishes to only one cycle every 24 hours at the poles. The

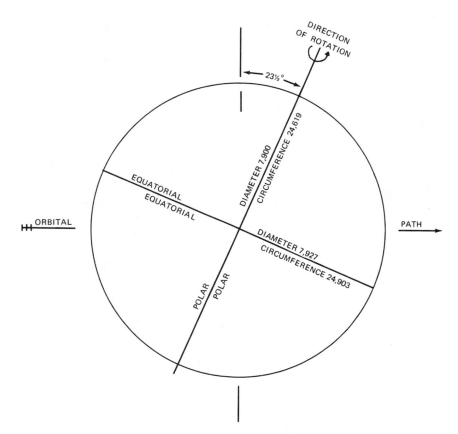

Figure 1.3. The earth's dimensions, its rotational direction and the tilt of its axis in relation to the plane of its orbit around the sun.

significance of this fact in determining the directional tendencies of motion in the atmosphere and the sea will become plain in a later chapter.

The sun's principal affect on the geosystem is three-fold: 1) it provides the gravitational center around which the earth pursues its elliptical orbit; 2) it is the earth's main source of light; and 3) it is the earth's primary source of heat. It is also the means of determining location accurately by latitude and longitude.

The rotating globe pursues an elliptical path entirely around the sun once every 365 days (actually 365 days and about 7 hours) at a distance that varies from a maximum of 94,800,000 miles in July to a minimum of 91,500,000 miles in January. The mean distance from the earth to the sun is thus nearly 93,000,000 miles. The earth's axis is inclined 23½ degrees from the perpendicular to the plane of its orbit about the sun. Thus through the course of a single year the portion of the earth exposed to the sun's rays is constantly changing (Figure 1.4). This continually shifting relationship proceeds in a most regular fashion within limits that can be specified with great accuracy.

It is the precision with which the earth orbits the sun, taking into account the earth's spherical shape and the 66½ degree tilt of its axis to the plane of its orbit, that has provided an accurate system for measuring *latitude*. Latitude is the distance north or south of the equator measured in degrees, minutes and seconds of arc. It was first established in the

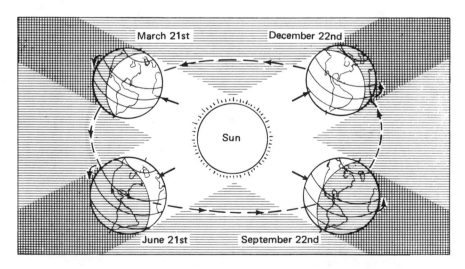

Figure 1.4. The portion of the earth's surface exposed to the sun's rays changes continuously, giving rise to variations in the intensity of global radiation according to latitude. (From U. S. Navy, **Meteorology for Naval Aviators,** Washington, 1958, p. 1-5).

4th century B.C. when scientists of Ionian Greece measured the angular distance between the poleward limits of the sun's position from year to year. It is convenient to speak of the sun's apparent motion across the heavens in a simplified exposition on the derivation of latitude. Greek astronomers observed that the distance between the sun's southernmost and northernmost positions, measured at the moment it attained its highest elevation above the southern horizon (at noon), was always 47 degrees (Figure 1.5). Midway between these limits they proposed an arbitrary theoretical line which they called the *equator* (from *aequus:*—equal). In this manner a basic reference line was introduced for the accurate measurement of position north or south of the equator anywhere between the poles. Between the equator and either pole the maximum angular distance is 90 degrees. Thus all latitudes lie between 0 and 90 degrees and are expressed as theoretical lines encircling the earth that are parallel to the equator, and hence are called *parallels* of latitude (Figure 1.6).

On about June 21st each year the sun reaches its northernmost position and begins its southward migration. This is the *summer solstice* (from *sol:*—sun; *sistere:*—to stand), and the mean position it attains at this time is represented on the earth's surface by the theoretical line called the Tropic of Cancer (from *trope:*—a turning). On about December 22nd each year the sun reaches its southernmost position and

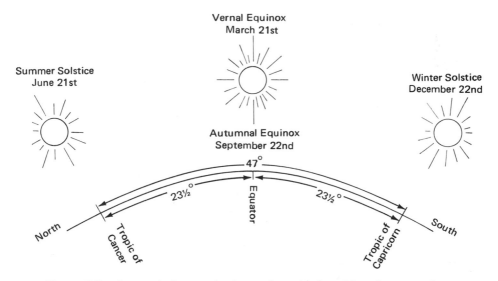

Figure 1.5. Seasonal changes in the sun's zenithal position. By measuring the sun's angular elevation above the horizon at the time the sun crosses the meridian for a period of several years, the position of the Tropics has been determined.

begins its northward migration. This is the winter solstice, and its mean position is indicated by the Tropic of Capricorn.

The earth's rotation on its axis provides the basis of measuring another fundamental dimension, *longitude*. This is distance east or west of an arbitrarily selected meridian, in degrees, minutes and seconds of arc. A *meridian* is a theoretical line passing through both poles and the zenith at the point of observation. Hence, every point on the globe possess a local meridian. In 1884, 25 nations convened at Washington, D. C. and agreed to establish the *Prime Meridian* at Greenwich, England. Longitude is now expressed as distance from the Prime Meridian and ranges in value from 0° to 180° east or west. The 180th meridian, approximately in the central Pacific, separates east longitude from west longitude.

The 180th meridian, throughout most of its length from pole to pole, is also used as the starting locus of each calendar day. The International Date Line coincides with it, except in the extreme northern and southern Pacific, through a common agreement among the world's leading nations. At any moment the date immediately west of the Date Line is one day later than the date immediately east of it, except when it is twelve noon at Greenwich.

Thus the spherical earth's motions within the harmonious mechanism of the solar procession provide a system for accurately indicating location on its surface. They also provide the precise program upon which are based the rhythm of the seasons and the alternation between daylight and darkness.

The seasons are mathematically determined according to the angular elevation of the sun (Figure 1.5). When it is precisely over the equator around March 21st, days and nights are equal everywhere on the globe. This event, called the *vernal equinox* (*aequus:*—equal; *nox:*—night), technically marks the start of spring (northern hemisphere). When the sun is over the Tropic of Cancer at the summer solstice, around June 21st, it marks the commencement of summer. Around September 22nd, the *autumnal equinox*, the sun is once more over the equator and days and nights are again equal throughout the world. At the winter solstice, around December 22nd, when the sun is over the Tropic of Capricorn, winter in the northern hemisphere formally sets in. South of the equator the seasons are reversed.

The duration of daylight and darkness also changes in a methodical manner as one season gives way to another throughout the year. The diurnal periods are accurately determined in hours, minutes and seconds by observation of the moment at which the sun's upper limb is tangent to the visible horizon at sunrise and sunset. At the equator daylight endures for 12 hours and 7 minutes on the date of each solstice. At latitude 40° the length of day on June 21st is 15 hours and 1 minute, but on December 22nd is only 9 hours and 20 minutes. At latitude 80° daylight is continuous at the summer solstice while darkness is continuous on December 22nd.

The changing length of day as well as the angular elevation of the sun can be measured in the course of a year with comparative exactness. Seasonal fluctuations of solar radiation intensity can also be calculated for specific latitudes with relative precision. Thus the effect of the earth's motions on incoming radiant energy from the sun provides a yearly rhythmic program of energy increase and decrease at the earth's surface. This is the energy that drives the mechanisms of the geosystem. The details of energy exchange are dealt with in Chapter 2. But the periodic precision of the earth's daily and yearly schedule of solar relationships is most profoundly modified by its geographical diversity. And the processes by which its continents, seas and atmosphere function together as a unified system reveal the consequences of this diversity in its fundamental geographical foundations.

GEOGRAPHICAL FOUNDATIONS

The surface diversity of the earth is a product both of its intrinsic materials and the forces acting upon them, giving them form, dimension and distribution. The earth's basic geography is the size, shape and distribution of its major land and water bodies. Unequal in size and shape, and unevenly distributed, they provide the foundation upon which all other geographic features continuously evolve.

To begin with, about 71 percent of the earth's surface is oceanic, 29 percent land. The comparative size of the world's principal land and ocean areas is shown in Table 1.1, as well as the proportions of land and sea on either side of the equator. It is evident from this that about 67 percent of the earth's land areas lie north of the equator (Figure 1.6). Thus in the earth's basic geography a fundamental imbalance exists. This alone affects most profoundly the dynamic operations of the entire geosystem. A sound grasp of the world's basic geography is indispensible to an understanding of all the interacting processes in that system.

How land and sea attained their present arrangement is very much a matter of dispute among geologists. It is sufficient here to state that the existing distribution has persisted since near the end of the Tertiary Period, about 1,000,000 years ago, when the general outlines of the continents and ocean basins had been formed. Since that time the forces of modification have been steadily at work altering terrestrial landscapes and submarine topography, through the constant exchange of energy, matter and momentum among land, sea and air.

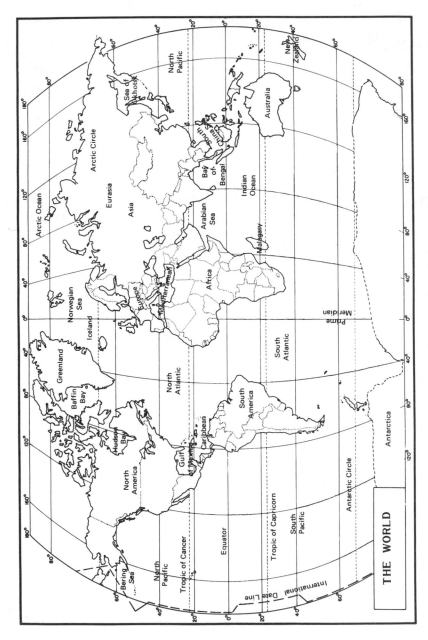

Figure 1.6. General geography of the world.

TABLE 1.1
Comparative Areas (square miles)

	Area	Percent
WORLD		
Total	196,940,000	100
Land	57,280,000	29
Northern Hemisphere 38,377,600		67
Southern Hemisphere 18,902,400		33
Sea	139,660,400	71
Northern Hemisphere 60,092,600		43
Southern Hemisphere 79,567,800		57
NORTHERN HEMISPHERE		
Total	98,470,200	100
Land 38,377,600		39
Sea 60,092,600		61
SOUTHERN HEMISPHERE		
Total	98,470,200	100
Land 18,902,400		19
Sea 79,567,800		81
OCEANS		
Pacific Ocean	70,000,000	
Atlantic Ocean	31,500,000	
Indian Ocean	28,375,000	
Arctic Ocean	5,540,000	
LAND MASSES		
Eurasia	20,910,000	
Asia	17,085,000	
Africa	11,685,000	
North America	9,420,000	
South America	6,870,000	
Antarctica	5,100,000	
Europe	3,825,000	
Australia	2,971,081	

The world's land surfaces have received most attention from geology through the historical study of the processes by which mineral and rock components form, combine and are modified. Surface deformation is the sum of all the changes occurring without let-up at the interface between land and air. By all the processes of gradation land surfaces are continually worn down in opposition to the constructive work of forces set in motion by adjustments of the earth's crust. But the major configurations of the land, the dominant surface features, endure with apparently ageless persistence. Their nature and distribution may thus be treated as relatively lasting features of the earth's present geography, compared with the more vigorous movements and speedier response to change of both sea and air.

MAJOR CLASSES OF LANDFORMS

In simplest terms there are four major classes of landforms: *plains, hills, plateaus* and *mountains*. The term *plain* in physical geography is employed to express the concept of flatness over reasonably large tracts of land, usually in excess of several hundred miles in area. The degree to which unrelieved flatness of terrain is approached varies widely. In many parts of the world, however, large expanses are known that are not only flat but virtually level as well. This is especially true of large *alluvial lowlands* where long-continued deposition has contributed deep, well-stratified accumulations of relatively fine materials. Levelling has proceeded through the sorting of particles by repeated flooding. Examples of extensive alluvial plains are seen in the lower Mississippi Valley, the Amazon lowland, the West Siberian plain and the North China plain. But most plains regions possess varying degrees of unevenness to which terms like rolling, undulating, rough or broken are applied. In all cases the absence of strongly contrasting relief features over most of the surface is characteristic.

The term *hill* is even less precisely definable than plain. On a broad plain of monotonous flatness a single low rise of only a few tens of feet may well merit the designation hill. Where an extensive lowland rises toward a bordering mountain region the intervening zone, the *piedmont*, is often spoken of as the *foothills*. Lacking an exact definition, the term is admittedly a relative one. Features of similar shape and dimensions may be known as hills in one part of the world and as low mountains in another. Where foothills leave off and mountains begin is frequently difficult to decide. Similarly an arbitrary choice must often be made between undulating plains and rolling hills. Furthermore, hills include a wide range of morphology and proportional dimensions. Most generally they are landforms in which the percentage of sloping land exceeds that of level land, and local relief, that is the vertical distance between hilltop and valley floor, is less than 1,000 feet.

A *plateau* is a broad, elevated area comparatively uniform in elevation. One or more sides rise conspicuously, usually more than 1,000 feet, above the adjoining countryside. Where the surface is uneven it is often termed a *hilly upland*. Where flatness predominates it is commonly called a *tableland*. In many cases mountain ranges tower above the mean elevation of the upland, and not infrequently appear in such profusion as to create a surface of *basin and range* topography. Much of Nevada, parts of Iran and northwestern Argentina possess this morphology of landscape.

The term *mountain* has been widely applied to a great variety of locally conspicuous surface features. But in a stricter sense a mountain is a prominent relief feature whose crown area is small compared with its base dimensions, having a high proportion of steep slopes on which a distinct climatic change is evident from base to summit. The vertical distance from summit to base is at least 2,000 feet and is usually much greater. Isolated peaks are often volcanic in origin, such as Mount Vesuvius or Mount Etna. An elongated series of peaks and ridges may be termed a range, as the Front Range of the Rockies. A complex of related ranges is frequently called a *mountain system*, like the Rocky Mountain system. An extensive distribution of separately definable systems occupying a large percentage of a continental area is commonly known as a *cordillera*. An example is the Andean cordillera of western South America, extending the entire length of the continent.

The four major classes of landforms—plains, hills, plateaus and mountains—are distributed in a highly irregular manner (Figure 1.7), and the percentage of each class varies from continent to continent. The distribution is far from balanced for the world as a whole. Plains account for about 45 percent, hills 15 percent, plateaus 5 percent and mountains 25 percent of the land surfaces on the planet. The remaining non-oceanic portions (10 percent) are the elevated ice-covered masses of Greenland and Antarctica. Clearly, plains are the most abundant surface features while plateaus are the least. Mountain terrain, occupying 25 percent of the land area, and plains about 45 percent, account for 70 percent of the total. Together they are by far the most significant land features in the mutual relationships among land, sea and air. This will become clear as the study of those relationships proceeds.

The most important fact of the earth's diversity is, without doubt, that its surface is about 71 percent oceanic. This means that most of the earth's atmosphere (about 71 percent) is at all times directly influenced by the various states of the sea. It is, of course, equally true that all oceanic surfaces are in turn under the constant influence of transient states of the overlying atmosphere. The interaction between sea and air is continuous. The fact that the earth's surface is dominantly oceanic thus indicates that the prevailing limits of atmospheric behavior are essentially prescribed by the physical properties of the sea. And it follows that the ruling planetary environment of the earth in space, with all its internal variations, is an environment dominated by the sea. This becomes clear upon considering the energy budget of the atmosphere.

REFERENCES

General
ESPENSHADE, E. B., (Ed.), *Goode's World Atlas*, 13th ed. Chicago: Rand, McNally and Co., 1970.

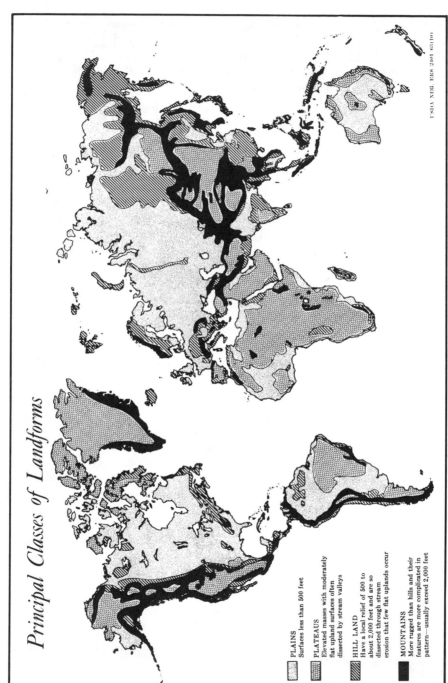

Principal Classes of Landforms

PLAINS
Surfaces less than 500 feet

PLATEAUS
Elevated masses with moderately
flat upland surfaces often
dissected by stream valleys

HILL LAND
Have a local relief of 500 to
about 2,000 feet and are so
dissected through stream
erosion that few flat uplands occur

MOUNTAINS
More rugged than hills and their
features are more complicated in
pattern—usually exceed 2,000 feet

USDA NEG. ERS 2401 621 101

Figure 1.7. Geographical distribution of the principal classes of landforms. (U. S. Department of Agriculture map).

GINSBURG, N., (Ed.), *Aldine University Atlas* Chicago: Aldine Publishing Co., 1969.
STRAHLER, A. N., *Physical Geography*, 3rd ed. New York: John Wiley and Sons, Inc., 1969.
Landforms
KING, L. C., *The Morphology of the Earth*, 2nd ed. New York: Hafner Publishing Co., 1967.
THORNBURY, W. D., *Regional Geomorphology of the United States* New York: John Wiley and Sons, Inc., 1965.
U. S. MILITARY ACADEMY, *Atlas of Landforms* New York: John Wiley and Sons, Inc., 1965.
WYCKOFF, J. *Rock, Time and Landforms* New York: Harper and Rowe, 1966.

Energy Budget of the Atmosphere

TOPICS

Composition of the atmosphere	Equation of energy exchange
Disposition of solar energy	Energy exchange processes

The atmosphere is the main integrating agent of the geosystem. It performs this function largely through the operations of its general circulation by which energy, matter and momentum are readily transferred from land to sea and from sea to land. In this, its great mobility is its chief advantage. Its mobility arises from its dominantly gaseous nature. It is easily placed in motion by the slightest imbalance in its density. Density differences arise mainly through differences in its temperature. Thus its capacity to absorb and release heat is most profoundly significant. Its heat exchange capacity results principally from the presence of water vapor, carbon dioxide and ozone. Dust also plays a significant but highly variable role.

COMPOSITION OF THE ATMOSPHERE

The enveloping mantle of the atmosphere is a mixture of invisible gases, minute dust particles and water vapor. Two gases alone account for 99 percent of the volume of dry air: nitrogen 78 percent and oxygen 21 percent. The remaining 1 percent is made up of exceedingly small amounts of argon, carbon dioxide, neon, helium, ozone and others. The percentage by volume of the atmosphere's main component gases (dry air) is shown in Table 2.1. These proportions appear to remain comparatively constant for the world as a whole from the surface upward through the first 15 miles. This is held true in the absence of sufficient proof to the contrary. For the same reason a balance is believed to exist between the addition and removal of the atmosphere's gases, although carbon dioxide is an exception as noted below.

TABLE 2.1

Chief Constituent Gases of the Atmosphere

(Dry Air, by volume, in percent)

Element	Percentage
Nitrogen	78.09
Oxygen	20.95
Argon	0.93
Carbon Dioxide	0.03
Neon	0.0018
Helium	0.000524
Krypton	0.0001
Hydrogen	0.00005
Xenon	0.000009
Ozone	0.000005

From National Aeronautics and Space Administration, *U. S. Standard Atmosphere,* 1962, Washington, D. C. 1962, p. 9.

The volume of water vapor in the atmosphere varies exceedingly from time to time and from place to place, for it is primarily a function of temperature. When air temperature increases, its moisture-retaining capacity is raised; when its temperature is decreased, its capacity is lowered. This relationship may be illustrated by considering a cubic meter of otherwise dry air at constant pressure, the temperature of which is gradually increased as shown in Table 2.2. Note that a temperature rise from 60°F to 80°F nearly doubles its moisture capacity. Because of its variability, atmospheric water vapor, even though in a gaseous state, is not included among the component gases of the atmosphere. In the later discussion of energy exchange it is treated as a separate, additional element. This is also true of dust. Amounts of each are highly variable in both time and place.

Atmospheric water vapor is supplied through the constant process of evaporation from bodies of water, from all moist surfaces, and from the transpiration of plants. The sea is the principal source; especially the oceanic regions of the lower latitudes upon which the sun's rays play from high overhead throughout the year. Sublimation from ice and snow also supplies atmospheric moisture in which the change of state from solid to liquid is accomplished without apparent liquefaction.

Atmospheric moisture is released through precipitation. Invisible water vapor condenses into cloud droplets on airborne particles having a strong affinity with water in a gaseous state. These are *condensation nuclei,* the most active of which are apparently sea salt particles and combustion effluents containing sulphurous and nitrous acids. Salt nuclei are extremely minute and usually range in size from 0.1 to 1.0 micron, although some are as large as 5 microns. (One micron is 1/1000 millimeter). They are abundantly present in the atmosphere, numbering up to 10,000 per cubic inch. Combustion nuclei are usually even smaller and their concentrations vary widely with distance from the source.

Water vapor tends to remain close to the earth's surface. About half the atmosphere's moisture content is estimated to lie below 6,000 feet, about 90 percent below 18,000 feet. Its presence is scarcely measurable at 6 or 7 miles above the earth (Table 2.3). For the world as a whole the moisture content of the atmosphere is thought to change only slightly with time. It is thus considered a conservative property of the global atmosphere, the balance being preserved between supply and depletion on a long term basis. A scattering of observations indicates, however, that water vapor amounts in polar regions are less than .5 percent of volume; in the middle latitudes they range from .5 percent in winter to 1.5 percent in summer; in tropical regions they reach 3 percent of volume. In the air over low latitude deserts water vapor amounts may reach 4 percent of volume. The scarcity of precipitation in desert localities is not for want of available atmospheric moisture, but is due to the tendency of desert air temperatures to remain well above the dew point, the temperature at which moisture in a given mass of air condenses.

TABLE 2.2
Moisture Content of
One Cubic Meter of Dry Air
at Various Temperatures

Temperature °F	Grams of Moisture
− 40	0.833
− 20	0.333
0	1.0
32	4.87
60	12.91
80	25.0
100	50.0

TABLE 2.3
Vertical Distribution of Water Vapor
in Percent of Volume in the
Mid-Latitudes

Height in Miles	Percent of Volume
5	0.05
4.5	0.08
4	0.13
3.5	0.19
3	0.29
2.5	0.37
2	0.46
1.5	0.63
1	0.80
0.5	1.06
Surface	1.30

Carbon dioxide is supplied to the atmosphere by emanations from volcanoes, gas wells, and springs, by the respiratory exhalation of plants and animals, and by the decay of organic matter. It is also supplied by the industrial processes of smelting, lime-burning, fermentation and the like. In addition, a major source is the combustion of wood and the fossil fuels created millions of years ago—coal, petroleum and natural gas. Carbon dioxide is mainly removed from the atmosphere by the process of photosynthesis, in which the green chlorophyll of plants converts carbon dioxide and water in the presence of sunlight into compounds of carbon, hydrogen and oxygen that resemble starches and sugars. In this process oxygen is released to the air. Carbon dioxide is also removed from the atmosphere through the formation of carbonates by organisms such as shellfish, and the weathering of calcareous rocks. Some carbon dioxide is also taken up directly by sea water. A volume of sea water ordinarily contains 60 times the amount of carbon dioxide found in an equal volume of dry air. Recent research suggests that in the Atlantic Ocean carbon dioxide is being absorbed north of the equator and south of 40° South, but is being released to the atmosphere between the equator and 40° South. There is also substantial evidence that the atmosphere in high latitudes contains much less CO_2 than the lower latitudes.

In recent years considerable concern has been expressed about the rising rate of atmospheric pollution resulting from the combustion of fossil fuels. Carbon dioxide is an important product of that combustion and the available evidence indicates that it has been increasing during the past 100 years. Around 1870 its atmospheric proportion was determined to be 290 parts per million while recent calculations indicate a present value of 330 ppm. A major cause of this apparent increase is the rising rate of fuel combustion for manufacturing, transportation and domestic heating purposes. Around large cities this is often conspicuously evident as *smog*. Here the high concentrations of human activity have led to a distinctly discernible increase in carbon dioxide and a decrease in oxygen.

Ozone is a form of oxygen (O_3) present in only trace amounts in the atmosphere (Table 2.1). Unlike carbon dioxide and ordinary oxygen (O_2) which are colorless and odorless, it is a faintly blue gas with a slightly acrid odor. It is introduced into the lower atmosphere by electrical discharges and by the decomposition of nitrogen oxide, a component of smoke from factory chimneys, forest fires and volcanoes. But it is in the upper atmosphere, at altitudes of 20 to 45 miles, that most ozone is formed by the photochemical action of near-ultraviolet radiation on ordinary oxygen. The process of ozone production is continuous and leads to a major concentration at altitudes of 15 to 20 miles. Below 15 miles ozone proportions tend to vary with the seasons and with transient weather conditions. It is transported downward by the vertical motions of atmospheric turbulence. Ozone effectively shields the earth from those portions of direct solar radiation having wavelengths shorter than .29 microns. Without its shielding effect the intensity of ultraviolet radiation would presumably be lethal for most forms of life on earth.

Minute solid particles, dry, rough, and collectively called dust, are present in variable quantities throughout the atmosphere. Mainly of terrestrial origin, they are supplied to the atmosphere by the smoke and ash of volcanoes, of industrial combustion, forest and grass fires, and by winds that sweep up mineral particles from desert regions and dry soils everywhere. Also included are pollen grains, fungi spores and bacteria, as well as salt particles from the sea. They range in size from 1 to 50 microns for the most part, although many are much smaller. Borne aloft by the motions of the atmosphere the finer components are held in suspension indefinitely returning to the earth's surface only when entrapped by falling precipitation. Over half the measurable dust of the atmosphere is found below 6000 feet above sea level, concentrations increasing toward the surface. Dust therefore adds to the density of the lower atmosphere and hence to its heat absorption and emission capacity near the surface. See Chapter 6 for further details.

ENERGY BUDGET OF THE ATMOSPHERE

The sun is the source of nearly all the energy, in the form of heat radiation, supplied to the atmosphere. Additional sources, such as other celestial bodies, the earth's interior, and the tidal effects of sun and moon, are only slightly significant and may be neglected. Tidal forces are dealt with in a later discussion of the sea.

The sun's surface temperature is apparently more than 10,000°F. Only about 1/5,000,000,000 of its estimated energy output is intercepted by the earth. This quantity is calculated to be almost two calories (1.94 cal.) per square centimeter per minute for a surface perpendicular to the solar beam outside the atmosphere. This value is called the *solar constant* and is also expressed as 1.94 langleys per square centimeter per minute. (A calorie is the amount of heat required to raise the temperature of one gram of pure water at a pressure of one atmosphere from 15° Celsius to 16°C. This is termed the *specific heat* of water). The intensity of solar radiation is not actually constant but varies from time to time as much as 5 percent. This is partly due to the appearance of darkened areas in the sun's outer gaseous region, the *photosphere*. These are more than 2,000°F cooler than the surrounding areas, and are called *sunspots*. The schedule of their variation in size and number is about eleven years and is known as the *sunspot cycle*. An annual rhythm of intensity variation also occurs in consequence of the earth's periodic change in distance from the sun. It will be recalled that the earth's elliptical orbit brings it about 3,000,000 miles nearer the sun in January than in July. Heat intensity varies with the square of the distance from the source. Hence, the solar radiation received in January is about 7 percent greater than that recevied in July. Considered on a long term basis, the solar energy received by the earth is balanced by an qual amount of energy reradiated from the earth into surrounding space.

The chief way in which radiant energy from the sun is taken up by the atmosphere is through transfer of heat from the earth's surface.

Solar radiation actually penetrating the atmosphere to the surface is called *insolation.* The angle at which the sun's rays strike the surface, called the *angle of incidence,* is determined by the sun's height in the sky, its angular elevation above the horizon. The sun is directly overhead, that is, its rays are perpendicular to a horizontal surface, at some time of the year only between the Tropics of Cancer and Capricorn. Because of the earth's sphericity the angle of incidence becomes increasingly acute toward the poles. At the Arctic and Antarctic circles it is never more than 47 degrees. Hence the sun's radiant energy at the surface in higher latitudes is diminished in two ways. First, by striking the surface at a more oblique angle the sun's beam is spread over a larger area, and second, it must pass through a greater thickness of the atmosphere before reaching the surface. Similarly, in the course of a single day solar radiation becomes increasingly effective as the sun's angular elevation in the sky increases between dawn and midday. Reaching a maximum when the sun is on the meridian at noon, it diminishes again toward sunset.

The average intensity of insolation for the entire year fluctuates systematically in consequence of the earth's motions, taking into account its sphericity and the 66½ degree inclination of its axis to the plane of its orbit around the sun. And the seasonal amounts of insolation, averaged in terms of langleys per minute, vary methodically according to latitude. Figure 2.1 is a diagram of the mean annual potential insolation curves

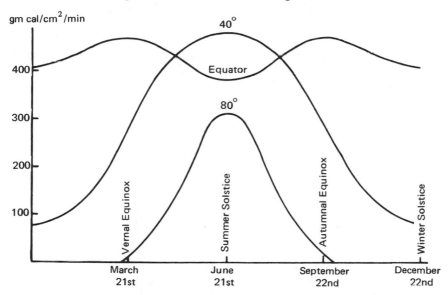

Figure 2.1. Mean annual potential insolation curves for the equator, for 40 degrees and for 80 degrees latitude.

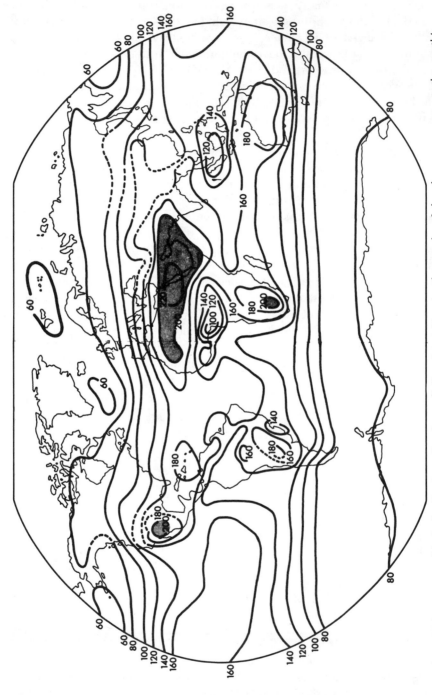

Figure 2.2. Geographical distribution of solar radiation intensities in thousands of langleys per year at the earth's surface.

for the equator, for 40 degrees and for 80 degrees latitude. Distance from the equator determines yearly amounts as well as seasonal fluctuations in potential insolation. Actual amounts of insolation delivered

to the surface differ seasonally and from year to year in response to random changes in solar output and transient atmospheric conditions. The geographical distribution of mean intensities for the year as a whole is presented on the map, Figure 2.2.

Insolation is greatly diminished by the amount and density of the cloud cover. About one-half the earth's surface is obscured at all times by clouds of varying thickness and translucence. The brilliant white upper surfaces of very dense clouds may reflect and scatter up to 80 percent of incoming radiation. This reflection is termed the *albedo*. Upper cloud surfaces reflect 24 percent of incoming solar radiation, the unclouded, gaseous atmosphere 7 percent, and the more substantial hydrosphere and lithosphere 4 percent. Thus the earth's total albedo is computed to be 35 percent, ranging between 30 percent in the low latitudes and 50 percent in the polar regions. The remaining 65 percent of incoming solar radiation is available for heating the atmosphere (Figure 2.3). These and other values of solar energy distribution are among the

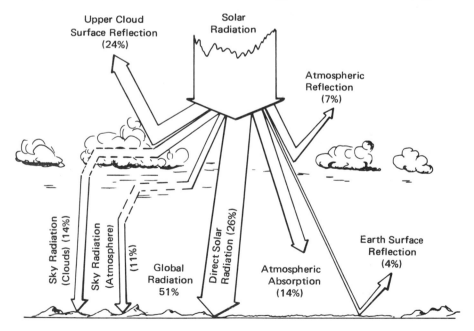

Figure 2.3. Disposition of incoming solar radiation. Global radiation is the sum of direct solar radiation (26%), sky radiation from the atmosphere (11%), and sky radiation from clouds (14%).

more recent of a series of calculations begun many decades ago and repeatedly revised.

Solar energy is transmitted to the earth in the form of electromagnetic radiation having wavelengths almost entirely confined between 0.15 and 4.0 microns (1/1000 millimeter; .001 mm). These limits are well within the shorter wavelengths of the thermal radiation spectrum which extends up to 100 microns (Figure 2.4). The atmosphere is transparent to most short wave solar radiation and is capable of absorbing only 14 percent of the sun's rays passing through it toward the earth's surface. Ozone, concentrated mainly in the upper atmosphere, absorbs nearly all the short wave radiation below 0.29 microns, creating a warm layer at the variable heights of the stratopause, between 25 and 35 miles above the earth in the middle latitudes (Figure 3.1). The rest of short wave radiation absorbed directly by the atmosphere is taken up chiefly by water molecules in the lower troposphere.

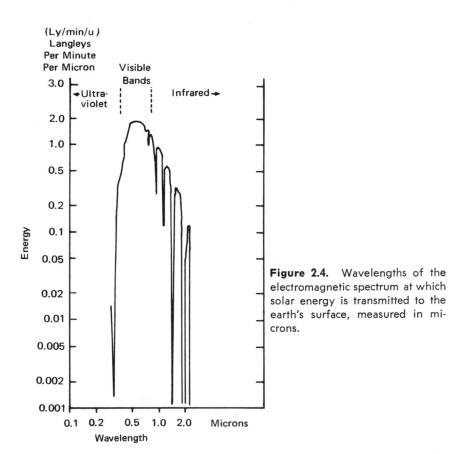

Figure 2.4. Wavelengths of the electromagnetic spectrum at which solar energy is transmitted to the earth's surface, measured in microns.

The remaining 51 percent (65 percent − 14 percent) of incoming short wave radiation reaches the earth's surface. Twenty-six percent is transmitted directly by the solar beam, and the rest is diffused sky radiation: 14 percent from clouds and 11 percent from atmospheric haze. Diffused sky radiation and direct solar radiation are together termed *global radiation*. At the earth's surface global radiation is converted to long wave thermal radiation ranging in wave length from 6.8 to 100 microns. Most outgoing terrestrial radiation is emitted at an average of 12.5 microns. To this longer wave thermal radiation from the surface the atmosphere is much more receptive, retaining it and delaying its return to space. This is similar to the way in which the glass panes of a greenhouse retain the warmth of the sun after its rays have entered, and for this reason is sometimes called the *greenhouse effect*.

The atmosphere derives most of its heat from the underlying surface of the earth. That it does so is most profoundly significant in the energy exchange relationships among land, sea and air. This will be more readily apparent in the discussion of atmospheric heat transfer below.

Water vapor accounts for 82 percent of the atmosphere's long wave heat absorption from below, while carbon dioxide accounts for 16 percent and ozone about 2 percent. Dust particles exert a highly variable effect and absorb, for the planet as a whole, only a trace of the long wave output. Once absorbed by the atmosphere it is then emitted as *counterradiation*. Also known as *back radiation*, this long wave emission takes place very near the earth's surface. Ninety percent of the counterradiation is estimated to occur within less than 110 meters of the surface. The atmosphere's heat capacity is thus greatest near the surface and diminishes upward as its density diminishes. However, measurable temperature fluctuations between day and night in the free air are detected up to altitudes of 1,000 meters. The heat-retaining capacity of the lower troposphere varies considerably from place to place and from time to time, becoming greater, for example, when moisture and suspended dust particles increase and less when the air is clean and dry. It follows that the much more rarified upper levels of the atmosphere are scarcely heated and remain constantly at very low temperatures.

The heat of global radiation (the 51 percent of incoming solar energy direct and diffuse) is disposed of in ways that differ according to the substance and temporal condition of the surface upon which it impinges. The variety of substances and the highly changeful condition of each make an accurate accounting of the heat exchange quite unrealistic. Table 2.4 shows heat absorptivity of contrasting surfaces.

Incoming radiation is treated one way on land and another way on water. Furthermore, the character of each major surface type is continually changing. Only a loose approximation of energy disposition is reasonable. But a generalized statement is necessary to understand the relative importance of energy exchange processes at the interface between the atmosphere and the underlying surfaces. Thus the energy exchange for the entire earth at the interface may be expressed tentatively in the following equation:

$$Q = T_1 + T_w + T_a + T_e + T_x$$

where Q = global radiation
T_1 = heat for warming land
T_w = heat for warming water
T_a = heat for direct warming of air
T_e = heat transfer to air by evaporation
      (latent heat of water vapor)
T_x = heat for such miscellaneous disposition as organic assimilation, melting ice and snow, chemical change and friction effects of wind.

Global radiation taken up at the interface between land and air is partly used to raise the temperature (T_1) of the solid materials forming

TARLE 2.4
Heat Absorptivity by Selected Surfaces
of Global Radiation (Percent)

Surface	Percent Absorbed
Oceanic water	98
Bays, lakes & rivers	90 – 95
Coniferous forest	90 – 95
Broadleaf forest	90
Green grassy meadows	85 – 90
Black dry soil	85 – 90
Dry light-colored sand	80 – 85
Dry grassy meadows	75 – 85
Dry plowed fields	75 – 80
Snow	10 – 20

TABLE 2.5
Depth of Heat Conduction
in Solid Materials*

	Rock	Wet Sand	Snow	Dry Sand
Daily (inches)	43	30	25	10
Annual (feet)	68	48	40	15

*After Geiger, R., *The Climate Near the Ground*, 4th ed., Cambridge, 1965, p. 33.

the surface. This it does mainly by conduction, and the depth of penetration is governed principally by the density of the material. The denser the material the greater its heat conductivity. Table 2.5 indicates the depths to which heat is conducted from the surface into solid rock, wet sand, snow, and dry sand, on a daily and an annual basis. But other conditions also affect the depth of penetration, such as color, texture, vegetation cover and slope. The possible combinations are infinite and constantly changing. However, the comparatively shallow depth of heat penetration in dry sand indicates that its surface temperature is considerably higher than that of wet sand. This is in fact born out in arid regions of the world. In northern Libya, for example, the surface temperatures of dry, loosely-shifting sand have attained values between 170° and 180°F.

Global radiation taken up by water bodies is principally used for evaporation. However, a portion (T_w) is also used to raise the temperature of the water mass. The processes are strikingly different from those involved in heating the solid materials of land. The heating of water proceeds very much more slowly. First of all the transparency of sea water allows short-wave radiation from both sun and sky to penetrate many meters below the surface. Ninety percent of the temperature increase in the uppermost meter of the sea is due to absorption of short wave radiation received directly from the sun. Daily temperature changes can be detected normally to a depth of about one meter. At 5 meters solar radiation accounts for only 50 percent, and at 10 meters only 25 percent of sea water temperature change. Long wave energy absorption and the process of evaporation are confined to a very thin film of water at the sea surface. Most of the total global radiation available to the sea is absorbed. Heat transmitted directly by the rays of the sun, however, is determined by the angular elevation of the sun. Assuming cloudless skies, the sea absorbs more than 95 percent of direct solar energy when the sun is at the zenith (90°); more than 90 percent at an elevation of 40° and more than 60 percent when the sun is only 10° above the horizon.

Despite the fact that most incoming radiation is readily taken up by the sea, the temperature of its upper-most meter fluctuates scarcely 1°F in 24 hours. However, fractional temperature changes have occasionally been observed to depths of 20 feet in 24 hours. Hence, temperature change proceeds only very slowly, not only from day to day, but from season to season as well. There are several reasons for this, arising from the nature of sea water and the enormous water masses involved.

To begin with, the specific heat of pure water is 1.0 (See page 28). This is greater than the specific heat of other substances. Iron, for example, has a specific heat of 0.11. This means that much less heat is required to raise the internal temperature of iron than of water. The amount of heat sufficient to increase the temperature of a gram of iron 18°F (10°C) would raise the temperature of a gram of pure water only a little less than 2°F (1.1°C). The specific heat of certain substances is given in Table 2.6. Note that the value for sea water with a salinity of 35 parts per thousand ($35°/_{oo}$) is 0.932, nearly equal to that of pure water since sea water of average salinity ($35°/_{oo}$) is 96.5 percent pure.

In addition, the dominant process of evaporation removes a portion of the surface water increasing its salinity and thus its density. Furthermore, evaporation cools the surface film of water. The denser, more saline water then sinks and is replaced by less saline water from below, within the upper few centimeters, in a process known as *thermo-haline convection*. The mixing action of surface wave motion, primarily induced by the friction effect of wind, also distributes heat downward from the surface. This is a process of turbulent convection in which warmer surface water is exchanged with cooler water below the surface. It is the most important means of heat transfer downward from the surface. This heat transfer effect reaches depths of 50 fathoms or more. The mass exchange of warmer surface water with cooler water from below is also accomplished by the process of *downwelling* which occurs along lines of surface convergence between water masses of contrasting density. Heat is also distributed horizontally in the flow of broad, persistent surface currents and the less perceptible movements of deep water masses.

By all the above processes sea water gains and distributes the heat of global radiation. Gaining and losing heat only very slowly, the world's oceans are the main heat reservoir of the entire planet. The slow rate

TABLE 2.6
Specific Heat of Selected Substances

Substance	Specific Heat
Water, distilled	1.00
Sea water 35°/$_{oo}$ Salinity	0.93
Wet mud	0.60
Ice	0.50
Wood	0.42
Sandy clay, 15 percent moisture	0.33
Limestone	0.22
Marble	0.21
Sandstone	0.21
Granite	0.19
Iron	0.11
Copper	0.09

From Smithsonian Institution, Misc. Coll., *Smithsonian Meteorlogical Tables*, 6th ed., Washington, D. C., 1951, pp. 405-406.

of internal temperature change acts as a planetary thermostatic control. In this, horizontal circulation is extremely important. On the continents lakes, rivers and lesser water bodies exert a similar, although a very minor thermodynamic influence of only local significance.

Global radiation for the warming of air (T_a) by exchange at the surface is accomplished mainly by evaporation. But it is also accomplished through conduction, convection and radiation. Conduction takes place from the monomollecular layer of the surface to the monomollecular layer of the overlying atmosphere with which it forms an interface.

When the surface is warmer than the air above, conduction produces expansion and uplift of the lower few millimeters of the atmosphere. This is compensated by the downward displacement of cooler elements, setting in motion an initial, small-scale convective exchange within the lower few centimeters of air. As long as the surface temperature remains higher than that of the air in contact with it, small-scale convection continues. Of greater significance is radiation. Long wave radiation of sensible heat is largely effective, as noted earlier, within the lower 100 meters or so of the atmosphere. Here the concentrations of water vapor, carbon dioxide and the suspended sediment of dust particles, selectively absorb energy of particular wavelengths in the electromagnetic spectrum. This long wave radiation, absorbed in daytime, is an important heat source for the earth's surface at night. It operates in response to the principle that a substance (in this case the lower troposphere), capable of absorbing radiation of certain wavelengths will reradiate energy of the same wavelengths at the same rate.

Carbon dioxide absorbs long wave radiation of wavelengths centering around three spectral values of 2.8, 4.3 and 14.9 microns. Water vapor is the most important absorbing agent, being receptive to wavelengths centered at 2.7 and 6.3 microns and above 12 microns. As electromagnetic radiation increases in wavelength beyond 12 microns the atmosphere becomes increasingly opaque and nearly all radiation is absorbed. Wavelengths around 4.0 and between 9 and 11 microns are not absorbed by the atmosphere regardless of how much water vapor or carbon dioxide it contains. These wave bands thus form "windows" through which incoming radiation is returned to space. Absorption spectra of the many types of dust particles in the atmosphere are not available. However, particle size and density of this highly variable atmospheric quality significantly affect the atmosphere's heat absorptivity.

Radiation is second to evaporation among the heat exchange processes. Conduction is effective only at the plane of contact between the air and the underlying surface. But conduction and radiation combine to heat the lower layers of the atmosphere from below. Convection provides the vertical exchange between portions of air warmed from below and cooled portions above. In this way heated air expands, rises and is replaced by cooler air. The vertical mass exchange occurs primarily because the atmosphere is heated mainly from below. But the process of mass exchange is not a simple one. It is complicated by horizontal as well as vertical motions of a highly turbulent nature. They are most frequently organized in the form of circulating eddies. For this reason the most important process of atmospheric heat transport is called *eddy diffusion*. It is sometimes termed dynamic convection, turbulent diffusion or simply turbulence. It is vital to the upward transport and cool-

ing of air to produce condensation. And the release of the latent heat
of water vapor by condensation (see below) is the chief way the atmosphere actually receives the heat derived from the earth's surface.

The vorticity of eddy diffusions develops in a wide variety of
scales. Turbulent mixing occurs both day and night, but the upward
transport of heat taken up at the surface proceeds mainly during the
day. Upward spiralling eddies, ranging in size from a few millimeters
to several meters in width, take shape, expand, disappear and reform
continuously as long as the requisite conditions prevail. These conditions are determined chiefly by the vertical temperature gradient, the
vertical moisture gradient and wind speed. On a very large scale, eddy
diffusion is vigorosuly active in the swirling vortex of every tornado,
of every large, mid-latitude cyclonic storm and every tropical hurricane. In all cases energy exchange and transport in kinetic form as
well as latent and sensible heat, are involved.

The study of eddy diffusion is one of the more important topics in
present atmospheric research. But the determination of actual eddy
diffusivity is extremely difficult. It may suffice here to state that the
vigor with which eddy diffusion proceeds becomes greater as atmospheric
instability increases. That is, the more humid the air, the steeper the
vertical thermal gradient and the higher the wind speed, the more
unstable it becomes. Upward velocity as a rule is about one-fifth of
horizontal wind speed.

The atmosphere is *stable* or *unstable* according to the degree to
which it is capable of maintaining vertical movement. It is stable when
upward or downward movement is inhibited. It is unstable when
vertical movement, once under way, continues indefinitely. Essential in
determining the degree of atmospheric instability at a given moment
are the *lapse rate* and the *adiabatic rate* of temperature change. The
lapse rate is determined by the temperatures at increasing altitudes at a
given moment in time. It is thus the atmosphere's thermal profile. It
varies considerably, but the *normal lapse rate* in the middle latitudes
averages 3.6°F for every change of 1000 feet in altitude. It becomes
very much steeper, that is, temperatures decrease much more rapidly
with increasing height, when the surface is heated excessively as on a
hot summer day. It is also steeper when unusually cold air moves in aloft.

Adiabatic temperature change takes place only within parcels of
air in vertical motion. No heat is exchanged with the ambient air
through which the parcel moves. It is an internal temperature change
arising from expansion or contraction of the volume of air. Rising, it
expands, its molecules collide less frequently, it becomes cooler. Descending, it contracts, its molecules collide more frequently, it becomes
warmer. Its expansion results from less atmospheric pressure as it rises,

and its contraction results from increased atmospheric pressure as it descends toward the surface. The rate at which unsaturated air increases or decreases in temperature is called the *dry adiabatic rate,* and is 5.5°F for every thousand feet of vertical displacement. Unlike lapse rate, the dry adiabatic rate is constant, as long as air is unsaturated. Humid, saturated air, on the other hand, changes temperature internally at the *wet adiabatic rate.* Moist adiabatic rates vary with the temperature of the air, ranging from 2° to 5.4°F per thousand feet. The lower value is typical of air at initially higher temperatures.

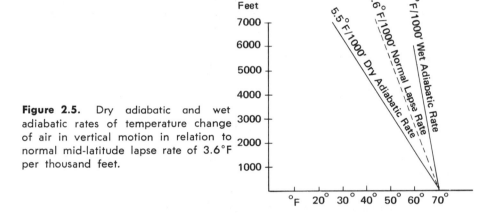

Figure 2.5. Dry adiabatic and wet adiabatic rates of temperature change of air in vertical motion in relation to normal mid-latitude lapse rate of 3.6°F per thousand feet.

Air is stable as long as the adiabatic rate exceeds the lapse rate. Thus, when a parcel of unsaturated air having a temperature near the surface of say 70° rises, its internal temperature will be 64.5° when it reaches an elevation 1000 feet higher. If the lapse rate is normal, the surrounding air at 1000 feet will have a temperature of 66.4°F. The upward moving air will be cooler than its surroundings at that level and will tend to return to lower levels. If, on the other hand, the lapse rate is greater than the fixed dry adiabatic rate, at perhaps 6°F per thousand feet, the temperature at 1000 feet will be 64°F and the rising parcel of air will hence be warmer than its surroundings and will continue to rise as long as existing conditions continue. Figure 2.5 illustrates the comparative values of dry and wet adiabatic rates under normal lapse rate conditions.

Over those parts of the world's land areas where evaporation plays only a seasonal role, chiefly outside the tropics, the surface exchange of heat to the atmosphere is accomplished mainly by radiation combined with small-scale interface conduction and the vertical diffusion of turbu-

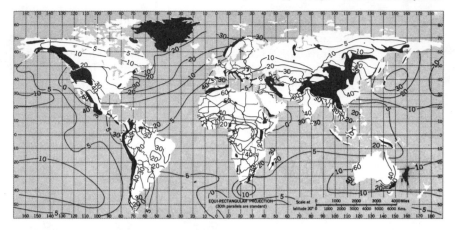

Figure 2.6. Geographical distribution of heat exchange intensities by non-evaporative processes. (After Trewartha, G. T., **An Introduction to Climate,** 4th Edition, New York, McGraw-Hill Book Co., 1968, p. 38).

lence. This is notably the case in arid regions. Figure 2.6 shows the world distribution of heat exchange by the combination of non-evaporative processes.

Evaporation is the principal process of energy exchange (T_e) at the earth's surface for the world as a whole. The main theatre of evaporation, as noted earlier, is the surface of the sea. According to calculations made by M. I. Budyko, an average of 88 percent of global radiation taken up by the sea is used for evaporation. The remaining 12 percent is used mainly to raise oceanic temperatures and, by radiation, to warm the air above the sea. Large amounts of heat are required to vaporize water, exactly how much depending on its initial temperature. At 0°C (32°F), 597 calories are needed, and at 30°C (86°F), 580 calories are needed to evaporate one gram of pure water. This amount of heat, removed from the surface film, lowers its temperature slightly. But it remains in the water vapor molecule and is borne aloft in the rising movements of turbulent air as *latent heat*. It is released to the surrounding air when invisible water vapor is condensed into cloud or fog or dew. Upon being released it warms the ambient air. The same amounts of heat are thus transferred to the air that were originally removed from the water surface. An estimated 80 percent of the atmosphere's fuel of operation is in the latent heat of water vapor. None is retained in the water of condensation. The very small amounts of heat gained by falling precipitation through compression and friction can be neglected in the overall heat budget.

The sea is the world's chief source of heat exchange by evaporation (T_e). Since perhaps 88 percent of incoming global radiation at the sea surface is used for evaporation and 71 percent of the earth's surface is oceanic, it is clear that evaporation is the principal heat exchange process for the entire earth. And the sea is by no means the only source. Also contributing to the evaporative process are lakes, streams, ponds, marshes and all moist surfaces on land, such as moist soil and rock. And a still more important source of water vapor on land is the moisture emitted by living plants. Water necessary to the metabolism of plant life is taken in at root level and released into the atmosphere through minute openings in the foliage called *stomata*. This process of water vapor release is *transpiration*. It is also sometimes called *productive evaporation*. The combined processes of transpiration from plants and evaporation from all other sources is sometimes called *evapotranspiration*. All in all, it is estimated that over 70 percent of global radiation is used for evaporation from all sources at the earth's surface into the atmosphere.

It is clear from the above that water in all forms plays a key role in the energy budget of the atmosphere. And while the atmosphere is the chief integrating agent of the geosystem, water by its dynamic function in heat exchange and transport is the indispensable substance through which integration is accomplished. Thus the earth's heat budget cannot be separated from the water budget. They must be considered together. Every change of state—solid, liquid or gas—participates in energy exchange and the movement of every gram of water is part of the heat transfer process. Eighty calories of heat are required to melt one gram of ice; 597 additional calories are required to vaporize one gram of water at 0°C. Upon condensing, each gram of water vapor releases 597 calories to the atmosphere, and upon freezing again each gram releases 80 calories.

It has been estimated that all atmospheric water vapor is turned over about once every 11 or 12 days, that is, evaporated from all sources and returned to the surface by precipitation. An estimated volume of 415,000 km³ per year is involved. The proportional distribution of the world's water supply at any given time is estimated to be: 1,370,000,000 km³ in the oceans; 23,000,000 km³ in the polar ice of the Arctic and Antarctic; 250,000 km³ in lakes and rivers, and 13,000 km³ in the atmosphere. During its residence time in the atmosphere water vapor is estimated to travel an average of more than 600 miles.

The main regions of continuous evaporation are the subtropical oceans as shown in Figure 2.7, where the moderate, steady flow of the surface trade winds, the predominantly sunny skies and persistently high temperatures are chiefly responsible for the process. Figure 2.7 thus shows those portions of the world where the release of converted solar energy at the surface is mainly concentrated. The latent heat of water vapor is later yielded to the atmosphere at higher levels by condensation, chiefly in the formation of clouds. Hence, where cloud formation and precipitation are virtually continuous, latent heat is given up to the atmosphere. This is a partial distribution of the surplus energy gained by the atmosphere in the lower middle latitudes.

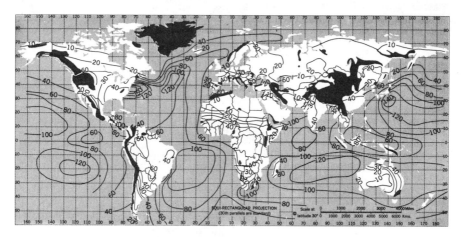

Figure 2.7. Geographical distribution of heat exchange by evaporation in one thousand gram calories per square centimeters per year. (After Trewartha, G. T., **An Introduction to Climate,** 4th Edition, New York, McGraw-Hill Book Co., 1968, p. 27).

The sea surface is not only the main source of the atmosphere's water vapor. It is also the source of the most important nuclei, minute salt particles, by which that water vapor is condensed and its latent heat released. Thus it is the vital medium that provides by exchange processes at the air/sea interface, not only most of the energy for heating the atmosphere, but also the essential raw material (water vapor) of latent heat dispersion, and the microscopic solids that expedite the release of latent heat to the atmosphere.

Evaporation from all sources proceeds as long as air is unsaturated. And the degree to which air is not completely saturated depends almost entirely upon its temperature. Suggested in Figure 2.7 is the importance of transpiration from terrestrial vegetation. Even in regions of heavy precipitation evaporation is powerfully effective. Thus in the tropical forests and in mid-latitude forests and grasslands during the growing seasons, transpiration is a major moisture source for the atmosphere. Figure 2.8 is a diagram of the proportional disposition of global radiation on dry land, on forested land during the growing season, and at sea. Values are approximate and are intended to indicate the relative importance of evaporation at the sea surface, evapotranspiration from humid, forested areas, and of sensible heat radiation from arid regions.

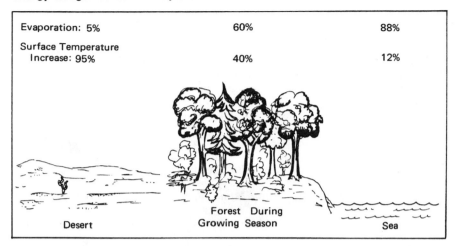

Evaporation: 5%   60%   88%

Surface Temperature
  Increase: 95%   40%   12%

Desert   Forest During Growing Season   Sea

Figure 2.8. Proportional disposition (in per cent) of global radiation on dry land, on forested land during the growing season, and at sea.

The amount of global radiation used for miscellaneous functions (T_x) is of negligible significance in the total energy exchange at the earth's surface. The main ways in which incoming radiation is disposed of are in heating the solid substances of land (T_1), heating water (T_w), direct heating of air (T_a) and by evaporation (T_e), of which the last function is the most important.

More than one-half the moisture supplied to the atmosphere by evaporation is released from the surface between latitudes 30°N and 30°S. In a more general context, the broad zone of the earth's surface between latitudes 40°N and 40°S provides the surplus heat of incoming radiation by all the processes of surface exchange. Areas poleward of 40°N and 40°S are regions of heat deficit. Large scale transfer of heat surplus to regions of heat deficit is the work of the systematic circulations in both the atmosphere and the sea. Of this work the atmosphere is credited with performing 75 percent and the sea 25 percent. Thus horizontal heat transport is profoundly important in preserving the balance of heat distribution on the earth. It is a product of the mobility of both air and sea.

Warm air from the lower latitudes constantly moves poleward into higher latitudes in the prevailing airstreams of the general circulation. The mobility of the atmosphere thus enables it to reduce the magnitude of strong thermal contrasts rapidly. The prolonged accumulation of either heat surplus or heat deficit is thus prevented. It is, on the other hand, the constant presence of thermal contrasts in the atmosphere that set up those geographical differences in density that in turn give rise to the flow patterns of the general circulation.

REFERENCES

BUDYKO, M. I., *Evaporation Under Natural Conditions,* (Translated from Russian), Jerusalem, Israel: Program for Scientific Translations, Ltd., 1963.

FLOHN, H., *Climate and Weather,* New York: McGraw-Hill Book Co., 1969.

GARSTANG, M., *Proceedings of the Sea-Air Interaction Conference,* Washington: U.S. Weather Bureau, 1965.

GEIGER, R., *The Climate Near the Ground,* 4th ed., Cambridge: Harvard University Press, 1965.

MALKUS, J. S., "Large scale interactions between sea and air,", in Hill, M. N. (Ed.), *The Sea,* New York: Interscience Publishers, 1962.

National Aeronautics and Space Administration, *U.S. Standard Atmosphere 1962,* Washington: 1962.

ROLL, H. U., *Physics of the Marine Atmosphere,* New York: Academic Press, 1965.

"Smithsonian Meteorological Tables", *Smithsonian Miscellaneous Collections,* Vol. 114, Washington: 1951.

U.S. Air Force, Cambridge Research Laboratories, *Handbook of Geophysics and Space Environments,* New York: McGraw-Hill Book Co., 1965.

CHAPTER

◆ 3 ◆

Structure and Circulation of the Atmosphere

TOPICS

Structure of the atmosphere   Subpolar lows
Equatorial trough   General circulation
Trade winds   Seasonal circulations
Subtropical highs

The general circulation of the atmosphere is the system of prevailing airstreams that transport energy, matter and momentum over the surface of the earth. Its integrating functions in the geosystem are performed almost entirely within the troposphere. The troposphere is the lowest layer in what is usually described as the structure of the atmosphere.

STRUCTURE OF THE ATMOSPHERE

The formless mass of the gaseous atmosphere is concentric with the more substantial lithosphere and hydrosphere which it surrounds completely, extending outward from the surface an estimated 600 to 700 miles. At present the structure of the atmosphere is deemed to consist of four distinct layers identified by temperature characteristics alone: *troposphere, stratosphere, mesosphere* and *thermosphere* (Figure 3.1). In simplest terms, temperatures decrease with increasing altitude in the troposphere, increase in the stratosphere, decrease again in the mesosphere and increase with altitude again in the thermosphere. Until recently it had long been customary to speak of three major atmospheric strata: troposphere, stratosphere and *ionosphere*. The mesosphere is the lower part of the ionosphere. Adoption of the four-layer concept of atmospheric structure is based upon the need for a realistic basis of distinction among the upper strata of the atmosphere. The upper limits of the three lower layers are identified by "pauses" (Greek *pauein*: to make

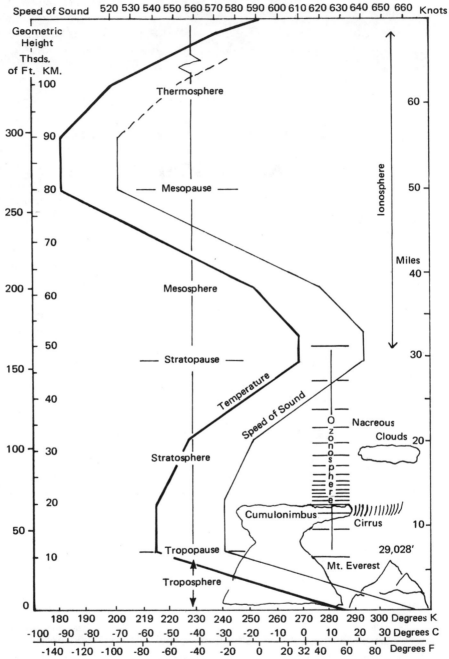

Figure 3.1. Vertical profile of the atmosphere showing the four-layered structure based upon temperature change with height. (After U. S. Navy, **Aerographer,** Washington, 1958, p. 20).

to change): *tropopause, stratopause* and *mesopause*. At each pause the qualifying temperature trend of each layer comes to an end and a new trend appears with changing height. Thus the tropopause is the base of the stratosphere and the ceiling of the troposphere. It is below this level that nearly all the atmospheric processes occur that are necessary to the complex interchange among land, sea and air.

Since the troposphere averages only between 5 and 10 miles in thickness, vertical motions in the atmosphere are very small in amplitude compared with the enormous range of horizontal motions. The distance from either pole to the equator is about 6000 miles, which is on the order of 1000 times the vertical distance from the surface upward to the mean height of the tropopause. The great mobility of the atmosphere permits the extraordinarily wide range of horizontal travel necessary to produce the essential exchanges by which the geosystem operates.

In the troposphere it is not only temperature that decreases with increasing altitude. So also do atmospheric pressure, density, dust and moisture content as well. Conversely, there is normally an increase with height in wind speed, transparency, and the intensity of solar radiation, particularly of ultraviolet radiation. Mean sea level pressure is 1013 millibars (29.92 inches), the weight of *one atmosphere* (Gr. *baros*: weight) as measured on the barometer. At about 5000 feet the pressure value is 850 mb, at 10,000 feet 700 mb, at 18,000 feet it is 500 mb and so on.

The rate at which temperature decreases with increasing height in the troposphere, as mentioned in Chapter 2, is a temperature change resulting from increasing distance from the primary source of atmospheric heat. The normal lapse rate for the middle latitudes of the northern hemisphere is 3.6°F for every 1000 feet of change in elevation. This normal temperature trend is often upset in the lower few hundred meters of the troposphere where surfaces are colder than the overlying air. Temperatures increase or at least hold steady where the air above is warmer, creating an *inversion*. Low level inversions are common throughout the year in the ice-covered polar regions, and frequently occur elsewhere as a result of nocturnal cooling of the lower few meters of the atmosphere. Such inversions are often a purely local occurrence and are known as *ground inversions*. A well-defined inversion of the tropics is the *trade wind inversion*, a most important feature of the tropical surface circulation (see below).

In the upper troposphere at levels where the normal lapse rate ends, at the tropopause, the stratosphere begins and for some distance upward temperatures remain isothermal, eventually rising in the upper stratosphere to values between 32° and 50°F at the stratopause. The tropopause is by no means a single, unbroken boundary layer. Instead there

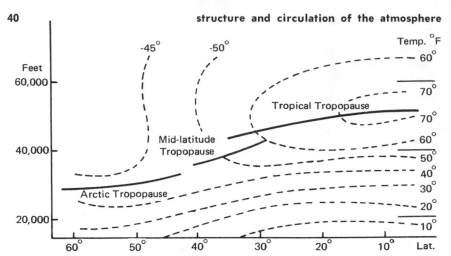

Figure 3.2. Relative positions of Arctic, mid-latitude and tropical tropopauses. (After Rumney, G. R., **Climatology and the World's Climates,** New York, The Macmillan Co., 1968, p. 22).

are several discontinuous and sometimes overlapping levels that increase in altitude from the polar regions toward the equator (Figure 3.2).

In general the multiple tropopause provides an undulating ceiling that varies in height from day to day throughout the year. It is usually just below this level that maximum wind speeds in the upper air are attained; where the jet stream appears. It is also the upper limit of turbulence, marking the boundary between the troposphere below, a region of much vertical air movement, and the stratosphere above, where most air movement is laminar and little turbulence is observed. Turbulence in the troposphere is one of its major distinguishing characteristics.

In brief, the incoming heat of global radiation, taken up at the surface of the tropical seas is chiefly converted by evaporation into the latent heat of water vapor. Raised aloft by the turbulence set in motion through heating from below, the moisture-laden air is brought to condensation point. This releases heat which augments the continued upward motions of turbulence, water vapor thus providing the essential fuel in the overall dynamic process. Turbulence in the lower latitudes is thus of first importance in maintaining the continuous operations of the general circulation. The most important feature of the tropical circulation is the convergence of the oceanic trade winds in the *equatorial trough.*

THE EQUATORIAL TROUGH—CONVERGENCE ZONE OF THE TRADE WINDS

The equatorial trough is a narrow, shifting zone of convergence between the northeast and southeast trade winds. It averages about 300

miles in width, although it expands from time to time to more than 600
miles. It is not an inflexibly persistent atmospheric feature, for it is frequently difficult to find and often disappears altogether. This is especially true over equatorial land areas. It is principally over the sea, between 5°S and 10°N that it is reasonably persistent. Figure 3.3 shows
its approximate alignment in January and July. The surface trade winds
draw together with diminishing speed in this zone of lower atmospheric
pressure, and for this reason the equatorial trough is also known as the
intertropical convergence zone, the intertropical convergence, or just
the *ITC.*

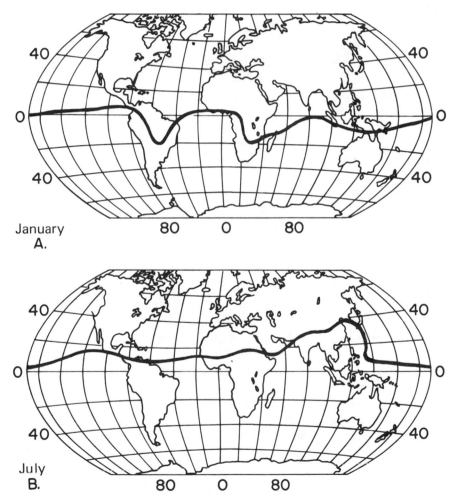

Figure 3.3. The equatorial trough (intertropical convergence). Approximate
alignment in January and July. (After Rumney, G. R., **Climatology and the
World's Climates,** New York, The Macmillan Co., 1968, p. 467).

The distinctive weather of the equatorial trough stands in sharp contrast to the gentle steadiness of the wind, abundant sunshine and scant precipitation of the trade wind zone (see below). Trough weather is typically changeful, in which alternate calms and variable winds, considerable cloudiness and abundant precipitation are characteristic. Rain occurs for the most part in the form of frequent showers and gusty squalls. Sudden, heavy rains, seldom lasting more than a half-hour, are commonplace during most of the year, falling from the dark gray bases of countless cumulonimbus clouds. Several thousand, (between 1500 and 5000,), individual precipitating clouds of this type are thought to be active in the equatorial trough at any moment in time. Among many scattered, smaller cumuli, the larger clouds are usually between two and three miles in diameter. They often grow vertically to more than ten miles in height, and rising beyond 50,000 feet, frequently penetrate the tropopause. These prominent cumulus towers of the equatorial trough perform the function of fuel pumps. Nearly 600 calories are available from every gram of water condensed. Energy at the rate of billions of calories per second is released to the middle and upper troposphere ($0.5 - 1.0 \times 10^{12}$ cal./sec.).[1] In the normal half-hour lifetime of each, the energy released is greater than that of a hydrogen bomb (@ 2.4×10^{14} cal.). This is the action that makes available to the atmosphere above most of the incoming global radiation received at the surface of the tropical seas. It is thus the initial action by which the general circulation is kept in motion.

As a rule the westward drift of clouds in the equatorial trough appears from below to proceed in random distribution. This is quite unlike the comparative order with which the ranks of snow-white cumulus are borne equatorward in the converging trades. However, most of the rainfall of trough weather is produced from cumulonimbus towers that are commonly arranged in the huge spiral distribution of systematic disturbances. These are usually 300 miles in diameter or more, and include anywhere from 50 to 150 precipitating cloud towers in each slowly gyrating system. It is believed that between 25 and 30 such disturbances are in operation continually, each of which lasts for a few days on its leisurely westward journey.

Air in the equatorial trough is typically unstratified and is remarkable for the great height to which its warm, humid properties extend. This condition predominates in the trough although it does not hold true throughout its length. Notable exceptions appear over the cooler oceanic surfaces of the eastern Pacific and Atlantic near the equator.

---

1. Malkus, J. S., "Large-Scale Interactions" in *The Sea*, ed. M. N. Hill (New York: Interscience Publishers, 1962), pp. 88-294.

The equatorial trough is considered the most homogeneous region of the troposphere, its uniform qualities reaching upward from the surface for 10 miles or more to the tropopause. The lapse rate here is much less than that for the middle latitudes, averaging only about 1.8°F per thousand feet of vertical change. For this reason the mechanical upthrust of the converging trades is chiefly responsible for inducing variations in temperature and moisture and the necessary turbulence required to set off local shower activity. When turbulence is vigorous enough, intensive squalls develop, and also a certain amount of support is given to large scale eddy disturbances.

The warm, humid, well-mixed surface air of the trades that lies below the drier air above the inversion, becomes deeper equatorward as the trade wind airstreams converge from northeast and southeast. Prevented by the sea surface from moving downward, they are forced upward. Thus the convergent uplift of an otherwise potentially stable atmosphere (lapse rate 1.8°F), is the mechanical means of raising the warm, highly humid air to the condensation point at higher altitudes.

Through the buoyancy of high-rising cumulus clouds in the upward displacement of latent and sensible heat, a mass exchange of air from the sea surface to the upper troposphere is accomplished. Far above the surface, from the middle troposphere to the tropopause, buoyantly billowing air spreads laterally, moving poleward into higher latitudes. Augmenting the volume of resident masses of air near the Tropics, air from higher altitudes accumulates over the subtropical oceanic regions to form relatively persistent high pressure systems. These are the *semipermanent subtropical oceanic highs.* They are the principal centers of mass dispersion for the general circulation of the entire earth. They are typically areas of virtually continuous subsidence and divergence in both hemispheres. It is from these centers of action that the trade winds, in their moderately steady flow toward the equatorial trough, close the vertical exchange cycle that originates in the equatorial trough.

The trade winds are the chief agents of evaporative activity at the sea surface. The latent heat acquired and transported into the buoyant turbulence of the equatorial trough is the major energy source for the atmosphere's motions. The trade wind zone occupies more than 31 per cent of the globe, between about 10 degrees and 25 degrees on both sides of the equator. The trade winds are the world's steadiest surface wind system, maintaining speeds of about 15 miles per hour in a persistent westward flow for several thousand miles across the tropical seas.

Unlike the equatorial trough the trade wind zone is typically stratified: a lower moist, homogeneous, tentatively stable convective layer, beneath a much drier upper troposphere. The moist layer is well mixed by the vertical exchange of constant turbulence and consists of a subcloud stratum to about 2000 feet, where it gives way to a cloud layer where small, buoyant cumuli, arranged in long lines or streets, rise to the base of the drier air above. In the moist layer the lapse rate is high (much higher than in the trough atmosphere), averaging about 5.4°F per thousand feet. Along the base of the drier upper air moisture decreases sharply and temperatures normally rise slightly to form a transition layer a few hundred meters thick called the *trade wind inversion*. The inversion inhibits further condensation and only a few billowing cumulus clouds extend their tops beyond it. The inversion layer determines the altitude reached by the cloud tops and toward the eastern limits of the trade zone, the outer trades, this ranges between 1500 and 4500 feet. Rising gradually southwestward (northern hemisphere) and northwestward (southern hemisphere) toward the unstratified equatorial trough, the inversion gains about 1000 feet in height every 500 miles, reaching about 12,000 feet toward the equatorial convergence. In the equatorial trough the inversion disappears.

The evaporative power of the trades is derived in part from their steadiness, in part through the increase of their own intrinsic temperatures and in part from the rising warmth of the underlying sea surface in the long trajectories they follow. Sea surface temperatures are high throughout the trade wind region, increasing very gradually westward from about 70°F to over 80°F. (Figure 3.4). Intrinsic air temperatures in the trades rise partly because of their passage over increasingly warmer seas, partly because of entering lower latitudes, and partly through the adiabatic warming of subsiding air. The unsaturated trade wind flow, moving equatorward from higher levels in the subtropical highs farther poleward, gain heat internally at the rate of 5.5°F for every thousand feet of descent. This is the *dry adiabatic rate* of dynamic temperature change mentioned in Chapter 2. The dry air of the trades produces precipitation on only two or three days per month. Thus the trade wind zone, principal source region of atmospheric moisture, is conspicuously deficient in precipitation. Daily weather characterstics are close to the long-range climatic means. In the moist layer, variations of temperature and moisture are very slight, and occasional synoptic disturbances are weakly developed. In a broader sense, the trade winds perform a coupling function between the turbulent, well-mixed moist lower troposphere and the uniform sea surface.

The trade wind convergence, set in motion by a gain in heat and moisture at the sun-warmed sea surface, is the initial atmospheric action of the general circulation. Air sent aloft by the cumulus towers in the score or more of organized disturbances exerts poleward pressure in both hemispheres. This takes the form of frequent but irregular pulsations at high level that send air into the higher latitudes where it cools

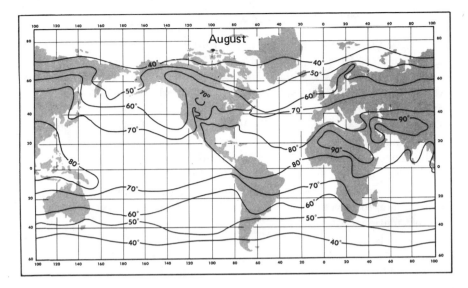

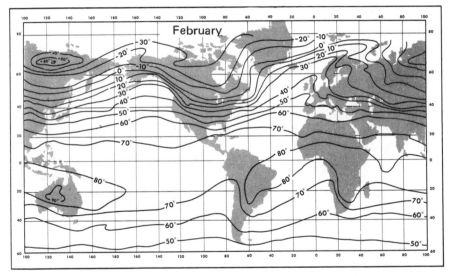

Figure 3.4. Geographical distribution of surface temperatures (°F) in February and August. (After U. S. Air Force, **Weather for Air Crews**, Washington, 1962, p. 2-7).

and slowly subsides to create the high pressure fields of the subtropics. At the surface these are persistent over the sea between 30 and 40 degrees of north and south latitude, and are known as the subtropical oceanic highs.

THE SUBTROPICAL OCEANIC HIGHS

The subtropical highs, continually replenished aloft by initial action in the equatorial trough, are major concentrations of air for driving the general circulation as a whole. They are regions where surplus accumulations of diverging and subsiding air repeatedly form, dissipate, and reform again to create semi-permanent circulations that are described as anticyclonic. Wherever, in the highly sensitive, fluid atmosphere, excessive masses develop, they press downward and outward in response to the tendency of all fluids to seek equilibrium. In the atmosphere such equilibrium is never attained, and a constant flux from areas of surplus to areas of deficit is the ruling condition. Hence, air perpetually moves from higher to lower density in the flow patterns that create the general circulation (Figure 3.4).

Wind is the gravitational response to density differences in the atmosphere that are commonly represented by differences in atmospheric pressure. Once in motion down the pressure gradient the actual direction of flow with respect to the earth's surface is determined by the combined influence of the Coriolis Effect, centrifugal force and friction. Bodies moving over the surface of the earth tend to proceed on a straight line. But the rotation of the spherical earth simultaneously shifts the directional grid system of meridians and parallels in relation to that motion is such a way to produce a curved path that causes moving bodies to be deflected toward the right in the northern hemisphere and toward the left in the southern hemisphere.

Opposing the curvilinear deflection of the Coriolis Effect (also called Coriolis Force, geostrophic effect, deflective force) is centrifugal force, and also friction, especially over rough terrain on land. At altitudes of 3,000 feet or more friction effects are virtually eliminated and the resultant wind flow is approximately parallel to the isobars that represent the levels of equal atmospheric pressure. Airflow at those heights above the friction layer is called *geostrophic* (Greek *ge:* earth; *strephein:* to turn).

Anticyclonic air movement in the northern hemisphere is clockwise and in the southern hemisphere is counterclockwise. Cyclonic circulations are the opposite. These are the rising vortices of lower atmospheric pressure into which air from the anticyclonic systems flows. Figure 3.5 illustrates the relationship schematically between high pressure and low pressure systems in the northern hemisphere. In anticyclonic circulations winds increase outward toward the perimeter from the central high pressure area of calm, and light, variable winds. The subsiding, diverging air, warmed by its descent, increases its moisture-retaining capacity. Thus in the central area clear, relatively cloudless skies prevail and precipitation is light and uncertain. Cyclonic circulations generally possess opposite characteristics. Winds increase toward the center (although in large, mature cyclones the central eye of the vortex is comparatively calm), relative humidity is high, systematic bands of cloud obscure the sky and precipitation increases inward from the perimeter.

From the major centers of action in the subtropical oceanic highs, warm, evaporative air of their anticyclonic circulations moves not only into the westward flow of the easterly trades but also into the east-flowing movement of the westerlies in the middle latitudes. Thus the oceanic highs are the chief source regions for the two main airstream systems in each hemisphere. But while the northeast and southeast trade winds are remarkable for their relatively gentle steadiness, the westerlies are known for their characteristic variability of both direction and speed.

The unevenness with which the eastward displacement of the mid-latitude westerlies proceeds is due to the unending conflict between air masses of contrasting qualities. The warm, moist air that repeatedly surges poleward from the subtropical highs meets colder air masses from the higher latitudes. Warmer air tends to override colder air at the surface, is cooled by uplift, and its moisture is condensed to form clouds and precipitation. As long as the influx of converging airstreams from contrasting air masses continues the precipitating disturbance is maintained. Such disturbances, in the form of barometric waves, troughs, or migrating cyclones, are the main sources of precipitation in the middle latitudes. Resulting from the convergence of air masses with contrasting properties, they stand in sharp contrast to the precipitating activity of the equatorial trough in which the convergent airstreams possess like properties. The upward displacement of humid, precipitating air in mid-latitude disturbances in which sea level atmospheric pressure decreases as the storm gathers strength, conveys heat by condensation to the upper troposphere. Divergent pressure aloft results, and air movement above the surface convergence is both poleward and equatorward. Mid-latitude depressions tend to move eastward along preferred trajectories and continue to function throughout the year as the mechanisms of upward energy transport and diffusion into the higher levels of the troposphere. They attain maximum intensity and frequency over the subpolar oceanic regions which are known as the *subpolar lows*.

THE SUBPOLAR LOWS

In the north Atlantic the *Iceland Low* extends for about 2500 miles northeastward from Newfoundland. In the north Pacific the *Aleutian Low*, again about 2500 miles in length, extends from near the Kamchatka

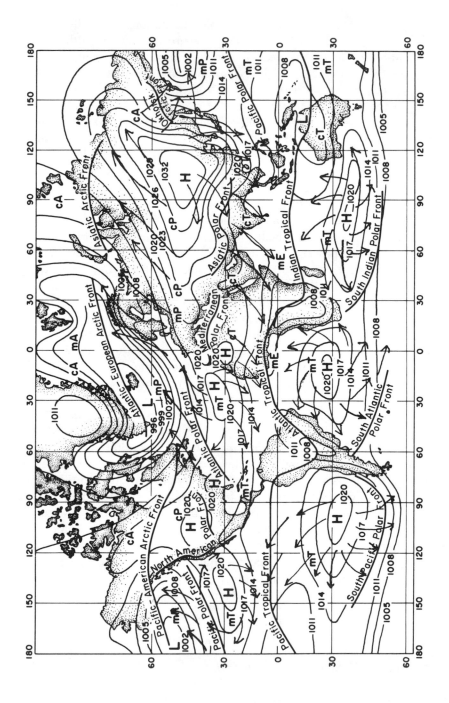

48

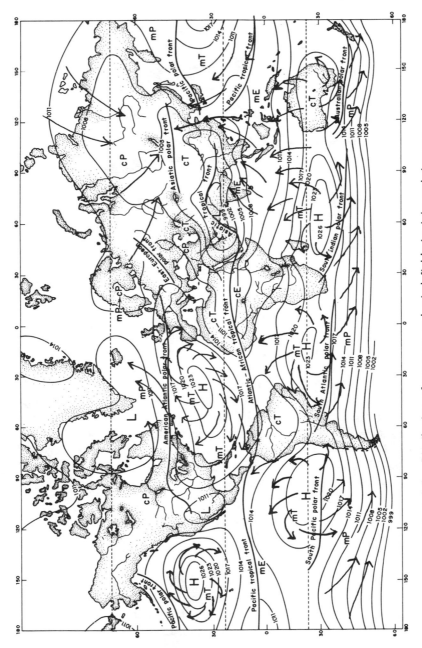

Figure 3.5. Distribution of pressure and wind fields in July and January.

49

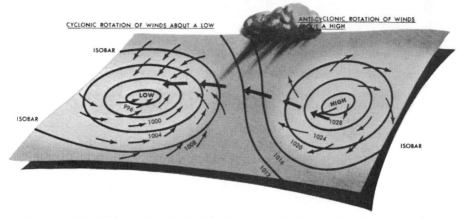

Figure 3.6. Schematic relationship between high and low pressure circulations in the northern hemisphere. South of the equator, wind directions are reversed. (From U. S. Air Force, **Weather for Air Crews,** Washington, 1962, p. 6-22).

Peninsula to the Gulf of Alaska. Surrounding Antarctica is a continuous belt of migrating depressions some 18,000 miles in length called the Roaring Forties. The oceanic subpolar lows are convergence regions, the product of frequent storm passage and intensification.

Mid-latitude disturbances contribute to the operations of the general circulation by providing the mechanism for transporting heat energy upward in the latent form of water vapor and releasing it through condensation at levels near the tropopause. Air that diverges above each disturbance contributes an increment to the resident high pressure regions of the subtropics and also to semi-permanent polar highs over the Arctic Ocean and Greenland and over Antarctica. Anticyclonic circulations from the polar highs produce weak and variable easterly winds, the *polar easterlies,* that recurve into the vortexes of migrating cyclonic disturbances over the subpolar seas.

THE GENERAL CIRCULATION

The meridional, north-south movements of the general circulation are diagrammed in profile in Figure 3.7. The figure seriously exaggerates the vertical dimension. It should be emphasized that most atmospheric transport is horizontal, and the scale of horizontal movement is on the order of 1000 times that of vertical displacement.

Mean upper airstreams, from about 18,000 feet to the tropopause, proceed, in very broad, sinusoidal sweeps. This is the mean configuration of horizonal transport at high level. The net displacement of all upper airflow outside the tropics is eastward in broadly sinuous circumpolar

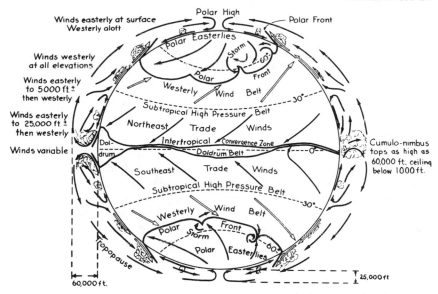

Figure 3.7. Schematic view of global wind systems and profile diagram of north-south (meridional) atmospheric circulations. (From U. S. Navy, **Meteorology for Naval Aviators,** Washington, 1958, p. 2-9).

whirls. However, deep meridional surges often occur sweeping far toward the equator on their western sides, and far poleward on their eastern sides.

Upper air movements frequently concentrate in very thin filaments of high speed momentum called the *jet streams.* The cyclones of the middle latitudes (large scale eddy diffusion) play a major role in maintaining the jet streams by the release of stored potential energy through the conflict between the contrasting air masses producing them.

SEASONAL SHIFTS IN THE GENERAL CIRCULATION

After the vernal equinox around March 21st the northern half of the earth becomes the summer hemisphere. The continents heat rapidly as the intensity of global radiation increases, raising the temperature of the air above them. Continental surface air expands, rises and enters the eastward sweep of the circumpolar flow aloft, inviting replacement at the surface by denser air masses from the sea. Humid oceanic airstreams advance over the margins of the major land masses with increasing frequency as the season progresses, although not everywhere with equal effect. Global radiation attains a maximum at the summer solstice around

June 22nd, and during the weeks that follow temperatures in the northern latitudes rise to a maximum for the year.

At this season all the main features of the general circulation shift northward. With the maximum intensity of solar radiation concentrated north of the equator the equatorial trough is displaced northward, reaching an extreme poleward position in August (Figure 3.3). Surface oceanic air from the southern hemisphere crosses the equator as the southeast trades respond to the change. In equatorial west Africa they recurve northeastward to bring copious volumes of warm, humid air to the Guinea Coast and continue eastward across Ethiopia. In the Indian Ocean the southeast trades pursue a sinusoidal course from northern Australia westward, then northward across the equator and finally northeastward to join the *southwest monsoon* of southern Asia. They also advance over the South China Sea to mainland China and converge with the northward flow of the northeast trades east of the Philippines where the equatorial trough assumes a north-south alignment from near Japan to the equator. Over southwest Asia a thermal low pressure field develops at this season, generally centered over southeastern Iran and western Pakistan. Surface airflow toward this summertime depression becomes highly heated and little effective precipitation reaches the earth. A similar thermal low appears over southwestern United States and northwestern Mexico, and here also inflowing air is highly heated and little effective precipitation results.

The subtropical oceanic highs intensify, expand, and in the northern hemisphere bring a southerly influx of warm, moist air to eastern portions of North America and Eurasia (Figure 3.5). To the western margins of those continents, however, south of latitude 35°, the equatorward flow of the anticyclonic circulation brings increasingly warmer, drier air. It gains heat both by subsidence and with diminishing latitude, through increasing global radiation. Furthermore, it is deflected away from land as it moves farther to sea to support the northeast trades.

In the subpolar Icelandic and Aleutian lows of the north Atlantic and Pacific fewer cyclonic disturbances converge with weaker intensities than those of the winter season, and the trajectories they follow lie farther northward. This is partly due to contraction of north polar ice on the sea surface which results from the warming trend in both sea and air during the northern summer.

South of the equator the northward shift of the subtropical oceanic highs and the expansion of south polar shelf ice over the seas surrounding Antarctica produce a northward displacement of the subpolar storm belt in the southern oceans. Fueled by the influx of warm, moist air from the oceanic highs, an endless succession of clockwise vortices progresses eastward and southeastward toward Antarctica over an almost unbroken oceanic surface. This is the season of maximum intensity for the furious turbulence of the disturbances that converge in the Roaring Forties and make this the largest region of continuous storminess on the globe. The frequency and vigor with which they form is chiefly the result of the very strongly contrasting deeply cold, dry air off the Antarctic plateau and the mild, moist flow of air from the subtropical seas farther north.

After the autumnal equinox around September 22nd, heat loss exceeds heat gain in the northern hemisphere. Frost and snow spread southward over North America and Eurasia while the dominantly oceanic southern hemisphere gradually enters the summer season of maximum warmth. The ruling features of the general circulation shift southward. The equatorial trough, although remaining near 5°N in the eastern Pacific and eastern Atlantic, bends sharply to cross the equator, reaching nearly 20°S in both South America and southern Africa. It also assumes a position south of the equator in the Indian Ocean, extending across northern Australia (Figure 3.3). The sinusoidal flow of air over the Indian Ocean is reversed by January. Advancing southwestward from southern Asia, the light winds of the northeast monsoon with far less strength than the southwest monsoon of July and August, cross the equator and curve eastward, then southeastward into northern Australia. By January northeast trade winds prevail for about two-thirds the distance around the globe between the shores of middle America and the east coast of Africa. Spanning the Atlantic, they sweep across the northern shores of South America toward the basin of the Amazon.

Over the rapidly cooling northern continents the atmosphere settles with increasing density as the deepening cold of winter progresses. Anticyclonic circulations predominate with the repeated formation of air masses that are typically stable, clear, dry and intensely cold. Clockwise circulations centered in northwestern North America and eastern Siberia dispatch surges of continental air toward the maritime margins and the adjoining seas.

South of the equator the oceanic highs supply warm, humid air to the continental low pressure systems that repeatedly form in northwestern Australia, and southern South America and Africa. The drift ice surrounding Antarctica contracts toward the shores of the polar continent. By March or April most oceanic ice has disappeared and the subpolar storm tracks approach the cold coast of Antarctica more closely.

The annual oscillations of the general circulation and the seasonal weather attending them produce the underlying rhythms in the processes that continually modify the world's geography. Systematic seasonal influences are exerted on both land and sea. In the dynamic integration of land, sea and air the work of the general circulation begins with the exchange of energy, matter and momentum between the atmosphere and the sea. Since the dominating surface of the earth is oceanic, it is necessary to consider the essential character of the sea.

REFERENCES

KENDREW, W. G., *Climatology*, 2nd ed., Oxford: Clarendon Press, 1957.
NEWELL, R. E., "The Circulation of the Upper Atmosphere," *Scientific American*, Vol. 210, No. 3, 1964.
PETTERSSEN, S., *Introduction to Meteorology*, 2nd ed., New York: McGraw-Hill Book Co., 1958.
RIEHL, H., *Introduction to the Atmosphere*, New York: McGraw-Hill Book Co., 1965.
TREWARTHA, G. T., *An Introduction to Climate*, 4th ed., New York: McGraw-Hill Book Co., 1968.
U.S. Navy, *Marine Climatic Atlas of the World*, Vol. VIII, Washington, 1968.

CHAPTER

4

The Sea
in the Geosystem

This chapter is devoted mainly to the circulatory functions of the sea in the geosystem. The circulation of the sea is dynamically coupled with the general circulation of the atmosphere. By its fluid character the sea possesses mobility that takes the form of waves and currents in response to a variety of extraneous forces. Atmospheric motion plays the leading role in maintaining the oceanic circulation through wind friction at the surface. Wind over the ocean imparts some of its momentum to the surface through *shear stress* or *turbulent drag*, and in this way loses much kinetic energy. Waves of varying magnitude are the immediate result of surface shear stress. On a long-term basis, however the world's prevailing wind systems bring about the large-scale transport of water masses and are thus the primary movers of the major ocean currents. Details of wave motion are discussed in the next chapter on coastal interaction.

The sea is the main theatre of activity in the geosystem, the atmosphere providing the integrating force. The sea is at once the chief source of the fuel (heat and moisture), essential to the functioning of the system, and the main recipient of its products. Incoming global radiation, taken up by evaporation at the subtropical sea surface, is supplied to the atmosphere to a large extent through the diffusive processes of the intertropical convergence. This is the atmosphere's primary fueling and pumping mechanism. Heat and moisture are taken up by the atmosphere in varying degrees over most of the globe, but the sea is the initial source.

By its general circulation the atmosphere works toward the equalization of temperature differences and provides a world-wide distribution of moisture originally derived from the sea. And the sea is the main source of condensation nuclei by which latent heat is released to the air, and by which moisture attains sufficient mass to fall as precipitation to the surface of the earth. Most moisture and salt crystals are returned directly to the sea by precipitation. Over ¾ of the world's yearly precipitation falls at sea. But of that which falls on land about 30 percent returns ultimately to the sea by surface runoff, and thus introduces into the geosystem an important secondary phase. The function of river discharge into the sea is to remove in solution and in suspension particles from the land that stands above the sea. These are the raw materials which mainly determine the salinity of the sea. Released from among them in the flying spray of surface turbulence are those very condensation nuclei by which precipitable moisture falls on the land to preserve the flow of stream systems into the sea and thus to preserve the continuity of the cyclical relationship between land and sea.

The salinity of the sea is one of the products of the geosystem. It is a valuable quality in determining sea water density and has many applications to biological problems. Salinity values are usually expressed in parts per thousand ($^o/_{oo}$) by weight, and are based upon determination of the chlorine content of samples of sea water. It will be seen from Table 4.1 that chlorine is the most abundant element dissolved in sea water. Since the proportions of all the other elements commonly found in the sea are believed to remain constant, the determination of chlorine gives a measure of the relative amounts of other elements. The salinity value thus provides a quantitative indication of the amounts of individual elements making up the chemistry of the sea in a given volume of sea water. Those elements are mainly present in a dissociated, ionic form.

SURFACE CIRCULATION OF THE SEA

The dominant circulation features of the sea are the surface currents forming enormous revolving *gyres* of approximately circular configuration that operate continuously in each major ocean basin (Figure 4.1). Integrated with the wind-driven surface flow are deep water counter currents and complex vertical motions arising from the dynamic processes of convergence and divergence. Subsurface circulation in the greater depths of the sea arises from density differences among deep water masses, movements responding to the density gradients, the rotation of the earth producing the Coriolis Effect and centrifugal force.

The revolving surface circulations are anticyclonic in response to the directional control of atmospheric movement—clockwise in the northern hemisphere, counterclockwise in the southern hemisphere. The general effect of their motions is to transport heat surplus poleward in

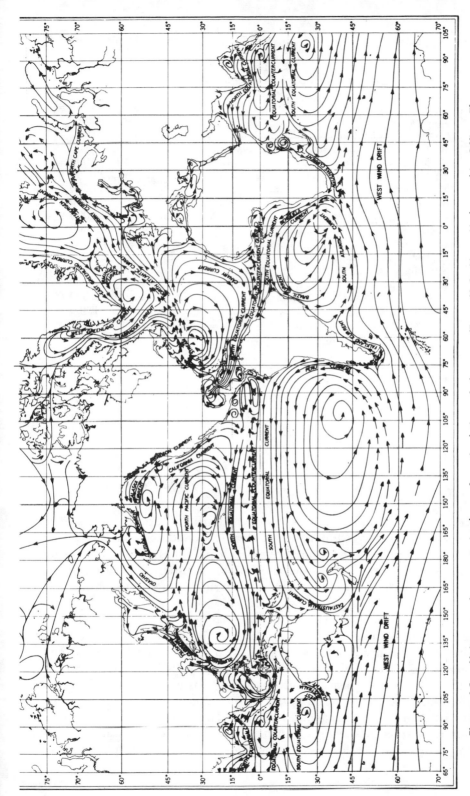

Figure 4.1. Major oceanic currents in the surface circulation of the sea. (From U. S. Navy Hydrographic Office, **The American Practical Navigator,** Washington, 1965, p. 720).

TABLE 4.1
Chemical Elements in Sea Water
at 35°/$_{oo}$ Salinity

Element	Milligrams per litre (mg/1)
Oxygen	889,000.000
Hydrogen	110,000.000
Chlorine	19,400.000
Sodium	10,500.000
Magnesium	1,300.000
Sulfur	904.000
Calcium	411.000
Potassium	392.000
Bromine	67.300
Carbon	28.000
Strontium	8.100
Boron	4.450
Silicon	2.900
Fluorine	1.300
Nitrogen	.670
Argon	.450
Lithium	.170
Rubidium	.120
Phosphorus	.088
Iodine	.064
Copper	.023
Barium	.021
Zinc	.011
Molybdenum	.010
Nickel	.007
Uranium	.003

From U. S. Naval Oceanographic Office, *Handbook of Oceanographic Tables 1966*, Washington, 1967.

western sectors and heat deficit equatorward in eastern sectors. The process is continuous but far from steady. Countless back eddies and counter currents of great complexity are created, plus a seasonal increase and decrease in the volume of water masses transported. And the entire circulating system experiences a meridional shift in the course of the year.

Many discrete types of water masses are involved, but there are two major kinds. Cold polar and subpolar water is of relatively low salinity but, because of its low temperature is very dense, and continually tends to subside into the depths. Warm, more saline tropical and subtropical water is much less dense because of its higher temperature, and tends to remain near the surface. Water masses from higher latitudes are colder due to the excess of heat loss over heat gain arising from the annual deficiency of incoming radiation. They are less saline due to a surplus of precipitation over evaporation. Low latitude water masses are more saline due to an excess of evaporation over precipitation, and are warmer through the surplus input of global radiation. These two major water mass types obey the basic principle that every water mass tends to move by the shortest vertical or horizontal path to that depth at which it will be in a stable equilibrium corresponding to its density. Differences in density usually create a slope called a density gradient, or *pycnocline*.

The initial drive of the anticyclonic gyre is the enormous volume of the equatorial current. This westward flow (Figure 4.1) is a consequence of the friction transfer of momentum from the dominant trades. These, in turn, trace their initial function to the concentration of heat uptake by evaporation from the oceanic surfaces where the highest intensities of global radiation are absorbed. Thus, incoming radiation, the primary energy source, provides both the kinetic energy of trade momentum which moves the equatorial current and the warmth which that current transports westward. Moving westward then poleward, surplus low latitude heat is moved into higher latitudes where it is released far from its source. This profoundly influences the limits within which thermal contrasts may develop for the world as a whole. The poleward flow of low latitude surface sea water provides warmth and moisture to the overlying atmosphere throughout the year. But its greatest period of influence is in winter when cold dry air from the continents or cold humid air from the subpolar seas moves out over the warmer surface. Cold air is warmed from below, more readily takes up moisture by evaporation, becomes unstable and more turbulent. The energy diffusion, set in motion by marked atmospheric contrasts, works toward the moderation of those contrasts and transfers surface oceanic warmth to the ambient air in the process.

The geographical distribution of sea surface temperatures is seriously modified by the operations of the oceanic gyres. This is reflected in the maps, Figure 3.4 of isotherms for August and February, the months in which seasonal extremes of thermal variation are experienced. The maps reveal the persistence with which the equatorward cool water transport continues. Along the continental shores past which these return currents flow, they are deflected seaward at intervals. This results in the upward replacement of surface water from below. Between the mainstream of equatorward movement are narrow, elongated, fluctuating stretches of even cooler water from the depths. They are generally confined to the subtropical latitudes of 35° to 20°, and are well-defined features of the west coastal waters off California, Chile, South Africa, northwest Africa, and to a slight extent off the west coast of Australia. The intensity of *upwelling* varies with the season, and appears to involve vertical movement within several hundred meters of the surface. The rate of upwelling is slow, and along the California coast proceeds at an average rate of about one meter per day. The process brings a steady supply of nutrients to the inshore surface waters, the enrichment of which supports a teeming population of micro-organisms and the larger animal forms that feed upon them.

Upwelling is usually in response to *divergence* in which horizontal movement occurs and with which it is coupled. Inshore divergence, however, allows only seaward lateral movement, and therefore tends to subside beneath adjacent water that is usually warmer, and of lower density. The subsidence of water beneath an adjacent, contrasting water mass is part of the process of *convergence,* again an exchange zone, normally very narrow and highly elongated, that is a common, widespread sea surface phenomenon. Convergence lines are frequently indicated by surface *rips,* areas of extremely agitated water and very often strong shear stresses between local currents of different strength and steadiness. Notable zones of convergence and divergence have been revealed in the waters surrounding Antarctica (Figure 4.2) and in the north Atlantic and Pacific.

In the north Atlantic the *North Equatorial Current,* primary component of the north Atlantic gyre, moves westward to enter the Caribbean Sea. Bifurcating among the Leeward Islands, it sends a branch northwestward past the Bahamas to augment the Gulf Stream. The Gulf Stream begins where the warm waters of the Gulf of Mexico emerge between Key West and Cuba. Bending sharply northward into

Figure 4.2. Surface oceanic circulation surrounding Antarctica. Mean positions of subtropical and antarctic convergences are shown.

the Florida Straits, it moves at a mean speed in excess of 3 knots, sometimes exceeding 4 knots. (A knot is a nautical mile (6,076 feet) per
hour.). The great volume of internal heat moving poleward is suggested
by the observed volume of water passing through the Florida Straits.
This amounts to 25 million cubic meters per second. Where it diverges
from the coast beyond Cape Hatteras, having gained volume through
merging with the Antilles Current, the mass of water in the Gulf Stream
has been calculated at some 82 million m^3/sec.

Curving northeastward the Gulf Stream is continuous with the
North Atlantic Current. This eastward flow divides as it approaches the
continent of Europe to send an arm into the Norwegian Sea, passing into
Arctic waters eastward as far as the ice-free harbor of Murmansk. Another arms curves southward to become the Canaries Current off northwest Africa. The cold Labrador Current extends southward out of
Baffin Bay to mingle south of Newfoundland in complex convolutions
with the warmer Gulf Stream. Over the shallow, cold waters of the
Grand Banks warm air off the Gulf Stream is chilled to form extensive,
persistent advection fogs. The contact zone between Gulf Stream water
and cold Labrador water is called the *cold wall* of discontinuity between
the two water masses. The Labrador Current and the cold air above
it prolong the life expectancy of icebergs originally detached from the
west coast of Greenland. Icebergs normally reach southward to latitude
40°, well to the south of Newfoundland, toward the end of every spring,
and occasionally have been seen near Bermuda and the Azores.

Between the north Atlantic gyre and the gyre of the south Atlantic
is a compensating *equatorial counter current* that entrains an arm of
the Canaries Current to enter the Gulf of Guinea south of West Africa.
The westward equatorial current of the south Atlantic gyre curves southwestward along the Brazilian coast as the Brazil Current, reaching poleward to about latitude 40 degrees. Here it meets the cold waters of the
Falkland Islands Current moving northward along the coast of Argentina.
The south Atlantic gyre contributes to the eastward flow of the *west
wind drift,* and then recurves northward in the cold waters of the Benguela Current which wash the shores of southwest Africa.

The west wind drift, perhaps better termed the *Antarctic circumpolar current,* integrates the circulations of all the world's oceans in a
continuous, unbroken flow, some 18,000 miles in length, that entirely
surrounds the icy plateau of Antarctica (Figure 4.2). It is more than
2,000 fathoms deep, and the enormous volume in continuous eastward
motion is estimated to be 110 million cubic meters per second. Essentially confined between latitudes 50° and 60°, it consists of several
distinct kinds of water that form convergences at the surface. The most
striking of these is the *Antarctic convergence.* This is a continuous,

narrow zone of discontinuity encircling Antarctica where very cold *Antarctic upper water* of low salinity from the melting of shelf ice and sea ice subsides beneath warmer but more saline water (Figure 4.3). The discontinuity is normally marked by a sharp thermal contrast of between 3° and 5°F. The dominant west winds in the converging circulation of migrating cyclones drive the Antarctic circumpolar current on its eastward course. The sea surface thermal discontinuity here intensifies the instability of the turbulent atmosphere, adding to the vigor of passing storm systems.

In the North Pacific the equatorial current proceeds almost due west for nearly 8,000 miles at an almost constant speed of from 1 to 1½ knots. Off the Philippines it divides into two main branches, one reversing into the equatorial counter current and the other heading northward into the *Kuroshio*. The warm Kuroshio, Pacific counterpart of the Gulf Stream, is broader, about 300 nautical miles wide, as it passes Formosa at between 1 and 1½ knots. It becomes narrower as it passes east of Japan, somewhat less than 100 nautical miles wide, and picks up speed to average between 2 and 2½ knots, although it slows again as it bends gradually eastward into the *North Pacific Current*. East of northern Japan it encounters the opposing flow of the *Oyashio*, a cold, less saline surface current from the Sea of Okhotsk and the Bering Sea. With the Oyashio it forms a turbulent convergence accompa-

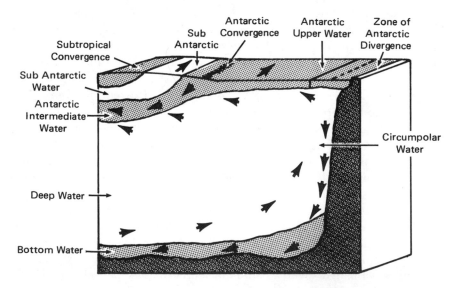

Figure 4.3. Block diagram of main water masses and convergence zones surrounding Antarctica. (From Nat. Sci. Fdn., **Antarctic Journal of the United States,** Vol. IV, No. 2, Mar-Apr 1969, p. 56).

nied by the formation of numerous back eddies above which appear frequent advective fogs. The north Pacific current moves eastward to separate into a counterclockwise circulation in the Gulf of Alaska and a southeastward flow that becomes the *California current*. This is a relatively cold surface flow that eventually turns southwestward to dissolve into the North Equatorial Current. Upwelling due to the persistent drag of northerly winds diverging from the California coast, is most vigorous in the summer season. The poleward reach of warmer water in the Gulf of Alaska is about ten degrees latitude less than in the Atlantic due to the barrier of the Aleutian ridge and the narrowness of Bering Strait.

In the south Pacific gyre the equatorial current advances steadily westward to pass north of New Guinea and reverse off the Philippines to join the equatorial counter current. Another arm, deflected into the Coral Sea, continues westward between Australia and the East Indies and through the Arafura Sea into the south equatorial current of the Indian Ocean. Southward from the Coral Sea the east Australian current flows into the Tasman Sea, curving eastward north of New Zealand to join the west wind drift. Approaching South America the Antarctic circumpolar current enters Drake Passage between Cape Horn and the Antarctic Peninsula. Here it is narrowly constricted and gains speed, up to 2 knots, before expanding once again on entering the south Atlantic. The north-flowing arm that becomes the Peru Current transports relatively cold water along the west coast of South America all the way to the equator. Around the Galapagos Islands cool surface water recurves westward into the south equatorial current.

The equatorial counter current of the central Pacific persists throughout the year. For the most part it becomes weak and quite narrow during the northern winter, but in summer shifts northward, between 5° and 10°N, and becomes wider and stronger. It averages about 300 miles wide in places, moves at about 1½ knots, and is continuous for nearly 8,000 miles from near the Philippines to the Central American coast. For part of this distance it is paralleled by a subsurface eastward flow called the *Cromwell current*, about 50 to 150 meters beneath the south equatorial current.

The Indian Ocean possesses a persistent anticyclonic gyre only in its major expanse south of the equator. North of the equator it is essentially reduced to the two broad embayments of the Arabian Sea and the Bay of Bengal. There the surface circulation responds to the strength and duration of the reversing winds of the south Asiatic monsoon. Alternating from the southwest monsoon of summer to the northeast monsoon of winter, the oceanic circulation reverses accordingly. By the end of April pressure and wind fields over southern Asia have adjusted to the increasing warmth of the sun's rays and by May the southwest monsoon begins. Between the *southwest monsoon current* that develops and the south equatorial current a marked convergence appears that persists across most of the ocean area except the extreme west. There the south equatorial current bends northward to become the *Somali current,* a relatively fast-moving flow, at times exceeding 4 knots, that induces upwelling off the island of Socotra where temperatures are frequently 10 degrees cooler (°F) than in the open sea. The coupling of surface airstreams and oceanic surface currents is nowhere better illustrated than here, for it is in just this western section of the Indian Ocean during the southwest monsoon that the strongest and most persistent airflow is observed. In July and August average wind speed is 30 to 35 miles per hour, and about one-third of the observed winds are of gale force or greater. Between 96 percent and 98 percent of the winds observed are from southwest to south.

Just north of Malagasy the south equatorial current is partly diverted southwestward to become the Mozambique current between Malagasy and the mainland, and this in turn is reinforced south of the island where it becomes the *Agulhas current.* This warm surface water moves close to the shoreline as it advances around the Cape of Good Hope. In so doing it opposes the eastward flow of the west wind drift and numerous vortexes develop of variable size and intensity. The warm Agulhas current gives way rather abruptly near Cape Town to the much cooler *Benguela current* that flows northward along the desert coast of southwest Africa.

SUBSURFACE CIRCULATION OF THE SEA

Below the surface, sea water masses move ponderously in slow response to their density differences. These arise chiefly from contrasting temperature and salinity characteristics. Among subsurface water masses temperature/salinity qualities tend to remain virtually constant. Their massive movements are thus remarkably steady, betraying very little variation with time, and discrete water masses are identifiable over great distances. Moving independently, they are interfoliated in well-defined strata. It is in Antarctic waters where the greatest complexity is observed.

Several distinct water masses are recognizable in the steady eastward set of the circumpolar Antarctic drift. In time these will doubtless be further subdivided, as study of this important region progresses. Its importance lies in its geographical relationship with all the oceans beyond it. Here, in the waters surrounding Antarctica, all the seas of the world are integrated in the circumpolar movement of oceanic strata. These are derived partly from the deep cold near the shores of the frozen

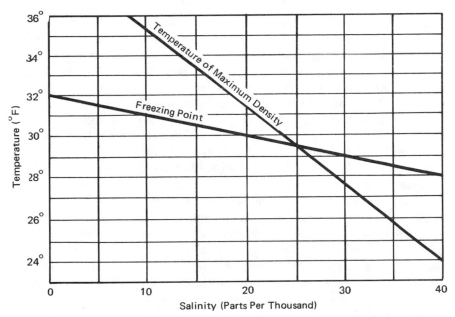

Figure 4.4. Temperature/salinity (T-S) diagram showing rates of density increase and freezing point decrease with diminishing temperature. (From U. S. Naval Oceanographic Office, **Handbook of Oceanographic Tables,** Washington, 1966, p. 22).

continent and partly from the warmer regions farther north in the Atlantic, the Indian and the Pacific Oceans. A meridional exchange in each ocean basin is continuous with the circumpolar set. Motions are intricately combined in the style of flat, slow-moving spirals or helixes in the eastward transport. This leads to multiple convergence and divergence as portions of the sea alternately rise and subside (Figure 4.3).

The relationship between sea water temperature and salinity and its density is of paramount importance in the formation of discrete water masses. From the temperature/salinity (T-S) diagram in Figure 4.4, it will be seen that fresh water (zero salinity) reaches maximum density at 39.4°F, several degrees above the freezing point. It therefore subsides to lower depths well before it might tend to freeze. As salinity increases to above 25°/₀₀ the freezing point of water becomes lower but at the same time its maximum density continues to be greater at above-freezing temperatures. Thus surface water subjected to progressive chilling continues to subside before freezing at salinities up to 25°/₀₀. When salinity becomes greater than this, however, surface water freezes and subsidence is arrested. In Antarctic waters the formation of sea ice is important in chilling and increasing the salinity of tremendous expanses of surface water.

The densest of all oceanic water is *Antarctic bottom water.* It originates chiefly during the winter season, June through August, when sea ice expands northward from Antarctic shores to latitude 60° or farther, adding to the south polar ice an area larger than the United States. Some 3,750,000 square miles of sea ice normally accumulated by the end of August. This adds to the persistent year round ice of the continental mass and adjoining shelf ice, having a combined area of 5,360,000 square miles. The year round ice of Antarctic accounts for about 90 percent of the world's total ice and 75 percent of its fresh water. Very cold water of low salinity in the winter season of expanded sea ice subsides toward the bottom, mainly in a downward settling zone in the Weddell Sea. This is a cold water *sink,* where a homogeneous, descending water mass prevails and spreads northward at great depth. In the Atlantic it has been identified across the equator, reaching perhaps as far as latitude 40°N, and in the Pacific as far as the Hawaiian rise, a volcanic mountain range in mid-ocean. The temperatures of the cold water sink in the high southern latitudes are as low as −2.2°C, about 28°F. The actual course of its northward movement is determined by bottom topography and seven broad thrusts have been found, three in the Pacific, two in the Atlantic and two in the Indian Ocean.

Above Antarctic bottom water is a layer of *deep water,* not as cold but more saline, the main movement of which is southward between north-flowing bottom water below and *Antarctic intermediate water* above. This layer originates mainly in the icy waters south of Greenland and also in the Sea of Okhotsk in the north Pacific. Antarctic intermediate water originates as cold, low salinity water at the surface that is warmed during the southern summer and then subsides along the Antarctic convergence. North of here it forms a layer between 100 and 250 meters thick beneath surface water from farther north that is both warmer and more saline.

FUNCTION OF OCEANIC CIRCULATION IN THE GEOSYSTEM

It is desirable to summarize the role of oceanic circulation in the operations of the geosystem. Its major importance is in operating toward the equalization of heat surplus and heat deficit for the earth as a whole. This it accomplishes by the poleward transport of warm currents and the equatorward transport of cold currents at the surface. Warm surface water becomes a source of both warmth and moisture for mid-latitude airstreams along the east coasts of the continents. Relatively cold surface water represses atmospheric moisture absorption, but chills the overlying air, producing frequent fog or low stratus cloud formations. Little precipitation results. This is in marked contrast to humid east coastal regions.

Desert conditions, in consequence, appear in the subtropical latitudes along the west coasts of the continents.

Where marked thermal discontinuities develop at the sea surface, atmospheric instability is enhanced, lending vigor to transient disturbances. The subpolar convergences of the north Atlantic, the north Pacific and Antarctic waters are unusually long and well developed. By their continual presence they account, in large measure, for the great intensity and frequency of subpolar migrating storm systems. And they mainly determine their trajectories as well. On a planetary basis oceanic circulation thus profoundly affects the distribution of precipitation. Its influence also extends to the distribution of relative humidity, of cloud cover, percent of possible sunshine, evaporative power of the atmosphere, wind speed and direction, and the incidence of storms and calms.

Oceanic circulation also proves to be a remarkably efficient system for the distribution of particulate matter entering the sea from the land. This is reflected in the map of sea surface salinity for the world ocean, Figure 4.5. The comparatively uniform chemistry of the sea is indicated here by the very slight differences in salinity from sea to sea. The key to the circulatory efficiency of the sea is the subpolar integration of all its waters in the steady eastward flow of the Antarctic circumpolar current. Suspended particles, not in solution, are less easily transported and are therefore not as well distributed. Of those that drift gradually downward and eventually become bottom deposits on the floor of the great ocean basins, most are the tiny fragments of solid material released by the metabolism of organisms in the open sea or the disintegrating remains of dead microorganisms. Bottom materials of these origins are called *biogenic*. Deep sediments derived from what were originally terrestrial deposits are termed *terrigenous*. Most bottom sediments are produced at sea.

In another way oceanic circulation extends the influence of polar ice toward the equator by transporting icebergs far from the point of their formation. In the north Pacific icebergs are normally observed to reach as far south as 42° latitude in the waters around Hokkaido, borne along by the cold Oyashio. In the north Atlantic icebergs are brought by the Labrador current well to the south of 40° latitude where they rapidly dissolve in the warm flow of the Gulf Stream. They have on rare occasions been observed as far south as Bermuda and the Azores. In Antarctic waters the flat-topped *tabular icebergs* that break off from the inshore shelf ice commonly advance equatorward beyond latitude 35°S in the cold waters of the Falkland Islands current off the coast of Argentina. And they reach similar latitudes near the Cape of Good Hope where they dissolve rapidly on meeting the warm flow of the Agulhas current. They also appear in the Great Australian Bight, around the same

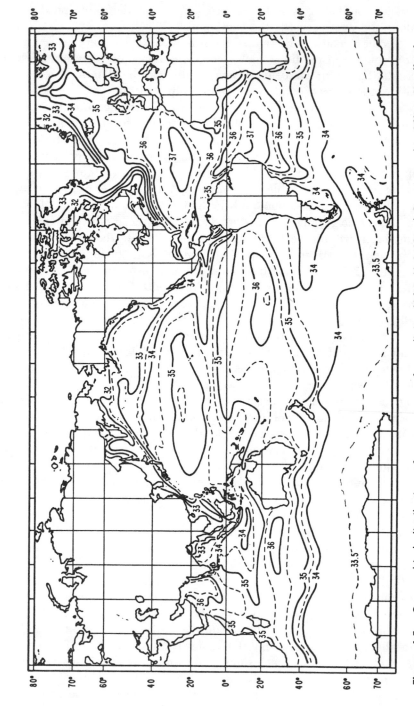

Figure 4.5. Geographical distribution of mean sea surface salinity. (After U. S. Naval Oceanographic Office, **Handbook of Oceanographic Tables,** Washington, 1966, p. 46).

latitudes. Nearer Antarctica the tabular bergs are often unusually large and are known to measure many miles in length. In February, 1953, a British whaling ship observed a tabular iceberg that measured 90 miles in length, 25 miles in width, and extended nearly 100 feet above the surface.

To the far-reaching effects of the oceanic circulation in the geosystem must be added the consequences of work performed by the sea directly on the margins of the land. In this, the processes of shoreline deformation are largely the result of waves and inshore currents and these are the subject of Chapter 5.

REFERENCES

CARSON, R. L., *The Sea Around Us,* New York: New American Library, 1951.
COKER, R. E., *This Great and Wide Sea,* New York: Harper Torch Books, 1954.
DIETRICH, G., *General Oceanography,* New York: Interscience Publishers, 1963.
DEFANT, A., *Physical Oceanography,* 2 Vol., New York: Pergamon Press, 1961.
HILL, M. N., (Ed.), *The Sea,* 2 Vol., New York: Interscience Publishers, 1962.
SHEPARD, F., *The Earth Beneath the Sea,* Rev. ed., Baltimore: The Johns Hopkins Press, 1967.
STOMMEL, H., *The Gulf Stream,* Berkeley: University of California Press, 1965.
VON ARX, W. S., *An Introduction to Physical Oceanography,* Reading, Mass.: Addison-Wesley Pub. Co., 1962.

5

Coastal Interaction
of Land, Sea and Air

TOPICS

Wind-driven waves and currents
Low-lying beaches
Irregular bedrock shorelines

Coral reefs
Tides and tidal currents
Abnormal events

The dynamic integration of land, sea and air is most strikingly evident along the shorelines where land meets sea. Here, where the interaction of land, sea and air is never at rest, the functions of the geosystem are performed within a time scale ranging from a few moments to many thousands of years. The ceaseless movement that proceeds from moment to moment, from day to day, from season to season and from year to year is revealed in the tangible features of coastal deformation. Evidence of unending interplay during ages long past is plainly displayed in the existence of raised beaches, buried forests, drowned stream valleys and submerged coral reefs. In the long range view, coastal deformation proceeds toward elimination of irregularities, toward making the shoreline smooth. It is prevented from reaching this goal by the constantly changing level of land in relation to that of the sea.

Most of the continental shorelines appear to have *submerged* in recent geological time due to a global rise in sea level following the last major glaciation. Clear-cut evidence of this is not everywhere easy to produce. However, where knolls and ridges of exposed bedrock have subsided, sea and land are strongly interpenetrated. Where deeply dissected mountains meet the water's edge, deep, elongated fiords penetrate far inland, alternating with the seaward thrust of bold, steep-sided peninsulas. This is the case along the fiord coast of Norway, of British Columbia and Alaska, of southern Chile and southern New Zealand.

Shorelines of *emergence* develop where land rises in relation to sea level, through broad-scale upward crustal movement. Along the California coast fragments of former beaches can be seen that now stand 150 feet or more above sea level. Along lengthy segments of the Atlantic coast at least four former beaches can be traced that now lie slightly inland and up to 40 feet above sea level, well removed from the reach of the sea.

WIND-DRIVEN WAVES AND CURRENTS

Coastal modification is brought about chiefly by the work of *wind-driven waves* and *longshore currents.* Their work is augmented at intervals by random events originating on land, at sea and in the atmosphere. The effectiveness of wave action as well as that of irregular processes is determined by the morphology of the shoreline, resistance of the component materials to movement, and the alignment of the coast to the direction of movement. Wind-driven waves and coastal currents expend both kinetic (related to motion) and potential (related to gravity) energy in performing the work of shoreline modification. In this way the energy of momentum is released that is gained from moving air through turbulent shear stress at the air/sea interface.

Essential to a discussion of surface wave motion are the standard dimensions (Figure 5.1) of length, height, velocity and period. Wave length is the distance from crest to crest; height is the vertical distance from trough to crest; velocity is the distance traveled in unit time; period is the time required for a wave to advance one wavelength. The dimensional characteristics of wind-driven surface waves are basically determined by wind speed, direction, duration and *fetch,* or distance traveled.

Surface wave motion is initiated when wind speed exceeds 0.7 meters per second. At starting speed surface tension ripples or *capillaries* that have a wavelength of less than 1.72 cm and reights of only a few millimeters appear on an initially undisturbed surface. In gusty weather they are frequently evident as transitory dark patches called *cat's paws.* They mark the initial stage of wave formation. From this beginning wave height rises in accordance with wind conditions to ten feet or more. Heights of more than fifty feet are comparatively rare.

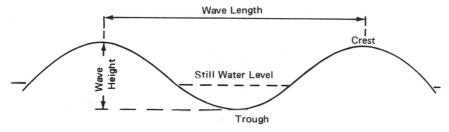

Figure 5.1. Standard wave dimensions of length and height.

To produce waves of about 65 feet in height in the open sea would require a constant wind of 42 knots for two full days with a fetch of more than 800 nautical miles. In December, 1922, the liner Majestic observed a wave about 88 feet high in the north Atlantic. As wind strength increases *wave steepness* also increases. Wave steepness is the ratio of wave height to length. The *critical steepness* of most waves is between 1:7 and 1:8 and becomes evident as their crests grow narrower and more translucent, and foam forms along the crest line. Once in motion, waves travel indefinitely until their energy is spent, either by gradual extinction in the open sea or by impact with the shore. Movement is outward in all directions, and waves from several distinct centers of wind action intersect to produce *cross waves*. As sea waves gain distance from the regions of initiating action they gain length and lose height to become *swells*.

Waves on the open sea are symmetrical oscillations of the surface that produce little forward advance of water. Water particles perform a circular motion (Figure 5.1) viewed in cross section, and thus pulsations of wave energy are transferred but not the substance of water itself. If water is deeper than half the wave length, the circular motion, dissipating downward, is unimpeded. Such waves are given the relative designation *deep water waves*. When water depth is about half the wavelength or less, the circular motion of particles meets with resistance at the bottom and waves are then called *shallow water waves*. Obviously the depth at which resistance is met is not constant and increases as wavelength increases. This variable depth (L/2) is referred to as *wave base*.

Shallow water wave particles are prevented from performing circular orbits and move in elliptical paths instead, their long axes more or less parallel to the inclined nearshore bottom. Approaching the shore they become strongly asymmetrical. Their troughs are long and flat, but their crests are sharp and steep. The portions of the nearshore bottom affected by incoming waves clearly vary according to wavelength, the slope of the underlying interface, and the variable depth of the sea as determined by the local tidal regime.

Most waves approach land obliquely, feeling the drag of the bottom progressively as each wave attains the depth of L/2. Wave height increases and length decreases, velocity diminishes and period lessens as each wave swings toward the shore. Waves rarely strike the strand head-on. Approaching land at an angle of, say, 45°, a wave bends by *refraction* to break against the shore at an angle of about 5°.

Among the many combinations determining the effectiveness of coastal interaction, three broadly distinct groups of landforms may be dealt with: 1) low-lying beaches of wave-washed sediment extending for considerable distances along shore between the outermost breakers and the inshore limits of wave action; 2) irregular shorelines of exposed bedrock; 3) coral reefs. Characteristic features of shoreline deformation are distinct for each group.

LOW-LYING BEACHES

The loose, unconsolidated material of which low-lying beaches are formed is mainly supplied by the outpouring of sediment-laden rivers entering the sea. Where incoming waves feel the drag of the bottom their work on this material begins. By dislodging and shifting the material they sort and deposit it according to the dynamics of their motions and the variety of particle size and volume. Very fine silt and clay particles are held in virtually continuous suspension; boulders, by contrast, are normally rolled back and forth along the bottom. Oscillating movement of small scale creates sand *ripples,* and beneath breaking waves of much larger scale low ridges or *bars* are formed that remain submerged, and shift position and change size as the state of the sea changes.

Wave height, increasing with diminishing depth, becomes sharply peaked and asymmetrical as soon as the depth is only about twice the distance from trough to crest. While its backside slopes gently seaward, its advancing front becomes steep, and when the depth is less than 1.29 x height (slightly more than one-quarter greater than height), it is unstable, collapses and forms a *breaker.* The consequent turbulence is mainly upward beneath a breaker, and bars form on the bottom along every breaker zone. Shoreward, successive lines of breakers may form, and beneath them a series of submerged bars and intervening troughs called the *surf zone.* Surf ends where the innermost breakers send sheets of water swiftly up the smooth incline that marks the limit of the beach (Figure 5.2). Here, along the shifting slope of upward*swash* and downward *backwash* the most vigorous interaction of land, sea and air on low-lying beaches is concentrated. This is the landward limit of action by wind-driven waves, where their energies are finally spent.

Solid material is kept in constant motion between the outermost breakers and the shore, and is drifted parallel to the shoreline by *longshore currents.* The rate of *longshore drift* is widely variable, but generally averages between 50 and 100 feet per hour. Slightly inshore, beyond the reach of normal wave action, finer particles of sand are subjected to more rapid displacement by the wind off the sea. In many parts of the world coastal *dunes,* piled high by the prevailing winds, take shape, become active, eventually stabilize, and at all times offer tangible proof of the constant interaction of land, sea and air. Where extensive

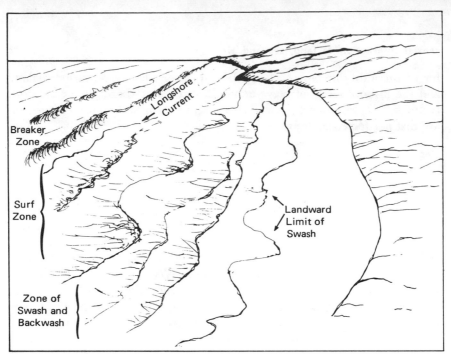

Figure 5.2. Features of wave action on low-lying beaches between the breaker zone and the landward limits of wave movement.

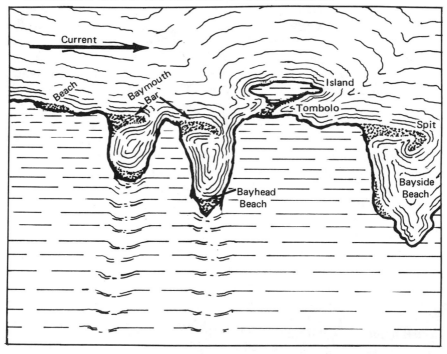

Figure 5.3. Geomorphic features of low-lying beaches produced by the action of wind-driven waves and currents. These modifications are transitory features of coastal interaction among land, sea and air.

beaches fringe the coast, where the supply of portable beach material is ample, waves and longshore currents unite to form shoreline features such as *spits, barrier beaches* and *islands, baymouth bars,* occasional *tombolas, capes* and *lagoons* (Figure 5.3).

IRREGULAR SHORELINES OF EXPOSED BEDROCK

Where the coast stands well above the sea and presents an elevated barrier of exposed bedrock to the crushing force of wind-driven waves, the interaction of land and sea differs strikingly from that of low-lying beaches. The enormous energy of a single wave is expended suddenly against the immovable rock. Wave energies have been measured at many points, and on the north coast of Scotland were found during one year to average more than one ton per square foot, increasing to about 3 tons/ft^2 in storms. At Dieppe, on the north coast of France, in February, 1933, a force of about seven tons per square foot was measured.

Yielding to the ceaseless work of dashing breakers, where every wave strikes with tremendous force, falls back and strikes again, bedrock is fractured at weak points, setting the stage for disintegration. Massive blocks, some weighing many tons, are dislodged and fall into the sea. Some become projectiles, and, hurled against the residual mass, aid the destructive force of the sea. Eventually *wave-cut cliffs* are formed, below which appear *wave-cut benches* or *platforms.* Large blocks of displaced bedrock are further fractured and disintegrated, accumulating at cliff base as rock debris, subsequently to be rounded by continued wave action into a *shingle* beach of rock fragments and coarse sand. Seaward of the rock platform loose material may accumulate to form a submerged *wave-built* terrace. Prominent headlands gradually wear away and recede, while intervening embayments slowly fill with sediment transported by the longshore currents and counter currents that work continuously to make the uneven shoreline smooth.

In the mid-latitudes wave action normally increases during the winter months with the increasing frequency and intensity of coastal storms. Under the conditions of more vigorous storm-driven wave action, much of the finer material is removed from beaches and the proportion of boulders and shingle is increased. During the summer season of less frequent storms the balance between coarse and fine material is restored. In the lower latitudes seasonal changes are very much less pronounced. But one feature of tropical shorelines has special relevance to the geosystem. This is the *coral reef*.

CORAL REEFS

The dynamic unity of the geosystem is nowhere better illustrated and nowhere more highly concentrated than on a coral reef. This is a strong, porous limestone structure created by marine organisms that extract the necessary chemical raw materials from the warm, turbulent sea. The primary reef builders are tiny reef coral polyps and minute, lime-secreting algae. They flourish where they can find support to which to attach themselves in clear, warm, well-circulated sea water penetrated by abundant sunlight. Sea water temperature is a vital limiting condition of their geographical distribution. They are mainly found where temperatures remain above 65°F, although many live where they drop to 61°F. Some corals in the Persian Gulf are known to withstand fluctuations between the low 90's and 52°F. Optimum temperature range is between 77° and 86°F. The requirement of ample sunlight limits the growing reef coral polyp to clear sea water no deeper than 150 feet. Other reef-building forms, however, live to depths of 300 feet, and live corals have been found at 580 feet.

The primary reef builders are the many species of coral polyp that extract calcium carbonate from sea water to build a hollow, tube-like shell enclosure around themselves, produce other polyps called buds which attach themselves to the shell of the first, grow their own calcareous casings, produce other buds and continue the process indefinitely. They are closely supported by lime-secreting algae that encrust the shells of other forms. Masses of skeletal remains from these and other marine organisms cemented together by calcium carbonate become the coral reef.

The quantities of nutrients and oxygen necessary for reef-building corals to flourish are supplied by the turbulent surf at the seaward edge of the reef. Required salinities are between 27°/oo and 40°/oo. Reef corals cannot live out of water and are thus not found above the low water limit of local tide ranges. But pounding seas and occasional storm waves generated by surface winds, dislodge fragments from the outer edge of the reef and distribute them unevenly behind it. Thus a surface is formed only a few feet above sea level as the reef continues to grow seaward. Coral reefs are constructive consequences of the unified operations of the geosystem. There are three main kinds: *fringing reef, barrier reef,* and *atoll.*

Fringing reefs are rough-surfaced platforms or benches that have been built outward from the land and are sometimes over a mile in breadth. Waikiki Beach on the south side of Oahu Island extends seaward about one-half mile. Barrier reefs are separated from land by elongated lagoons and channels that are often many miles wide. Numerous examples are found among the mountainous islands of the western Pacific. The largest coral area in the world is the Great Barrier Reef off the northeast coast of Australia. Stretching northward for about 1500 miles from near Brisbane it is an elongated complex of reefs and low islands nearly 100 miles wide. Most of the world's coral reefs are found in the central and western Pacific, with lesser numbers in the Indian Ocean and among the islands of the West Indies. Atolls are the most common coral reefs. Hundreds of them are scattered over a distance of some 6,000 miles between the Tuamotus and the Carolines. An atoll is a series of very low, very narrow islands more or less encircling a shallow lagoon. Most of them occupy a surface of only a few square miles, but some are more than 20 miles across and over 200 square miles in area.

Pacific Ocean atolls have formed where the tapering peaks of submarine volcanoes provide the support necessary for coral reef growth. Many thousands of volcanoes have risen from the floor of the Pacific. According to one explanation, a volcanic cone that has broken surface at its summit begins to subside from the weight of its own mass on the resilient lithosphere. Meanwhile its exposed peak has been subjected to erosive attack by sea and air, and to it reef corals have attached themselves to begin the work of reef building. Where erosion has flattened a volcanic peak an atoll may form (Figure 5.4). This is possible when the upward rate of reef building is equal to the subsidence rate of the supporting mass. The string of organically constructed islets grows, wave-washed debris of skeletal fragments accumulates on the floor of the encircled lagoon, and the entire atoll-forming process continues. Providing its volcanic foundation neither subsides nor emerges at a rate faster than that of reef formation, the relationship continues indefinitely. In 1952, holes were drilled in Eniwetok Atoll, Marshall Islands, at two points to 4600 and 4200 feet, before encountering lava. Eniwetok is about 25 miles long and 12 miles wide.

Some volcanic peaks do not become flattened and acquire fringing or barrier reefs. The flat-topped summit is required for the foundation of a coral reef atoll. Many submarine volcanoes subside beneath the waves to remain submerged and form *seamounts*. Those that are flat-topped are known as *guyots*.

Most atolls are between 2 and 5 feet above sea level. A small number of islets are somewhat higher than the average, suggesting subsequent emergence. Global sea level fluctuations are without doubt significant in the relative vertical displacement of volcanic peaks and sea surface. Wind-drifted dunes of calcareous sand are sometimes formed on larger reef foundations and are several times higher than the normal surface of reef debris.

The atoll is a unique synthesis of the geosystem, revealing the dynamic interplay of wind-driven sea and sunlight with the organisms that build solid rock. The atoll is itself a dynamic system. In exquisitely delicate balance between the living, lime-secreting organisms and the sunlit sea, it is tentatively poised on the tapering tip of a large, submerged

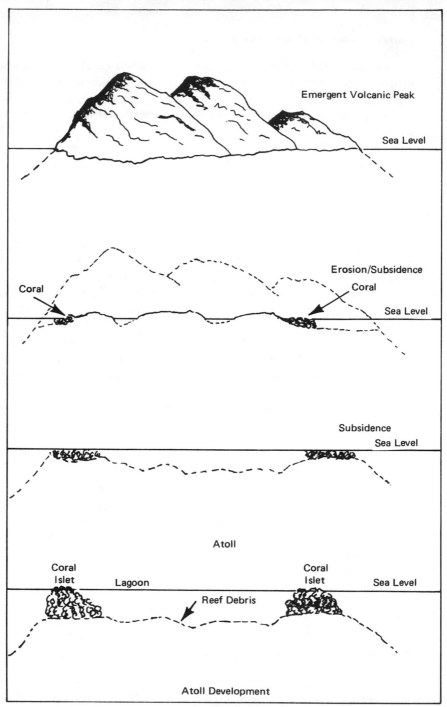

Figure 5.4. Development of a coral reef atoll on a submerged volcanic peak. The atoll is a unique synthesis of the geosystem, a product of the dynamic interplay of wind-driven sea and sunlight with the organisms that build solid rock, tentatively poised on the tip of a large, submerged mountain rising several miles above the ocean floor.

mountain. The soft-bodied polyp, extracting calcium and carbon dioxide, secreting calcium carbonate in a self-enclosing tubular shell, producing more polyps which do the same in the warm, saline surf, creates a limestone platform with which sea and air continue to react to support a minute ecology of progressive change.

TIDES AND TIDAL CURRENTS

The work of wind-driven waves and currents is augmented by the tide—the periodic rise and fall of the sea and the resultant tidal currents. The earth's tides, in the sea, in the atmosphere, and apparently also in the plastic lithosphere, are produced by the gravitational attraction of the sun and moon. The moon is the chief tide-producing agent. Gravitational attraction between two bodies is in proportion to the product of their individual masses and inversely proportional to the square of the distance between them. Although the sun's mass is enormously greater than that of the moon its effect is less than half that of the moon, for its distance is about 390 times farther than the distance to the moon (about 93,000,000 miles vs. about 240,000 miles).

Since gravitational attraction increases with diminishing distance, the maximum pull of the moon is centered around the point on the earth's surface nearest it (Figure 5.5). Opposing this attraction is the earth's gravity, drawing all objects at the earth's surface toward its center. The resultant attraction is called *tractive force* and follows the curvature of the earth's surface toward the point of maximum lunar pull. It is tractive force that causes the fluid sea to move toward that point, creating a slight bulge at the sea surface. This is the broad, low, major crest of the tidal wave that rises against the land as high tide. On the opposite side of the earth the surface is farthest from the moon and, therefore, least affected by its gravitational pull. Again the mobility of the sea allows tractive forces to move it toward the point of least attraction. The main mass of the earth is nearer the moon and is pulled away from the point farthest removed, producing at the sea surface a second, lower bulge, that also rises to bring high tide to the coast. In this way the moon establishes at the earth's surface two persistent long waves consisting of two slight bulges and two intervening depressions. The earth's frictional drag pulls the tidal bulges ahead of the moon's zenithal position. The earth, rotating once every 24 hours (actually every 23 hours, 56 minutes and 4 seconds), moves its surface beneath these waves. It is thus the earth's rotation that delivers oceanic tides in a regular, periodic schedule to the coasts of the continents and island shores all over the planet. The program of tidal rise and fall proceeds with such precision that these events can be predicted years in advance with great

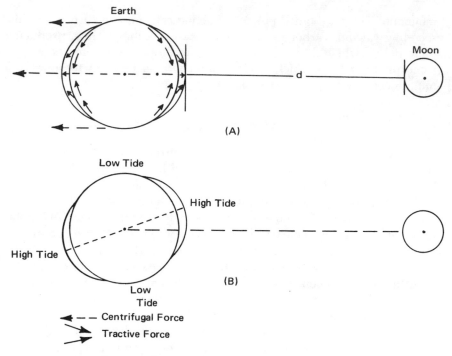

Earth
(A)
Moon
d

Low Tide
High Tide
High Tide
Low Tide
(B)

◄ ── Centrifugal Force
➤ Tractive Force

Figure 5.5. Tide-producing effect of the moon on the fluid sea. Tractive forces are chiefly responsible for producing the extremely long, low wave of the tide. Frictional drag of the solid earth on the sea carries the outward bulge of high tide ahead of the moon's position.

accuracy for all parts of the world, subject to modification by random atmospheric effects.

The sun modifies the moon's primary effectiveness. When sun, moon and earth are nearly in line (Figure 5.6) at either new moon or full moon, they are said to be in *syzygy,* and the sun's gravitational pull is added to that of the moon. At such times high tides are higher, and low tides are lower than mean tide levels, and are then called *spring tides.* When the moon is at right angles to a hypothetical line between the sun and the earth, at both the time of first quarter and third quarter (Figure 5.6), it is said to be in *quadrature.* Tidal amplitudes are less than mean tidal range, and are called *neap tides.*

Throughout most of the world two complete tide cycles occur in slightly more than 24 hours, although there is much variation of this basic cycle, and some areas experience only one tide per day. The local *tidal range,* vertical displacement of the sea surface between high and low tides, varies widely also, depending partly on the configuration of the shoreline. Bay of Fundy tides are the greatest observed. At the entrance the range is 9.5 feet, but at the head of Minas Basin, some 200

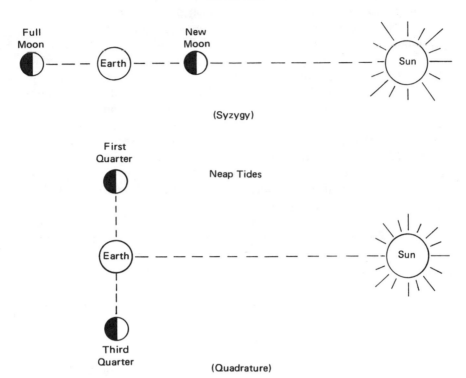

Figure 5.6. Spring tides and neap tides. When sun, moon and earth are nearly in line (syzygy), tidal ranges are greater than normal (spring tides). When the moon's position is about at right angles to the alignment of earth and sun (quadrature), tidal ranges are less than normal (neap tides).

miles up the bay, the normal spring tide range is 50.5 feet. In the Gulf of Mexico, by contrast, normal ranges are only slightly over two feet, and in places there is only a single tide per day.

The vertical movements of the tide set in motion currents of great power and complexity among the intricate channels and waterways of the shoreline. Moving large quantities of particulate matter in suspension and coarser material along the bottom, they continually alter the bathymetric (bottom) features of inshore waters. *Flood tide* occurs when the sea rises toward *slack water* at *high tide*, and *ebb tide* occurs until slack water is again reached at *low tide*. Ebb tide currents are usually stronger, especially in estuaries where receding sea water is supplemented by fresh water of rivers entering from the land. Tidal currents increase in velocity when the sea moves through narrow constrictions. The seaward flow of the East River in New York often exceeds

6 knots, and in the Straits of Georgia between the mainland and Vancouver Island, the tidal current often reaches 10 knots.

Tidal currents transport and deposit suspended material, sorting it out as it is released according to changes in current turbulence and speed, and the buoyancy of the particles. They also accomplish the exchange of chemical properties between inshore waters and the open sea. They further act to equalize temperature differences during successive tidal cycles. When water is moved from the depths into the shallows it becomes warmer in summer, returning seaward at the surface on the ebb tide. In winter surface water becomes colder in the shallows. On the shallower tidal flats salt-tolerant grasses may take hold to form *tidal marshes* that tend to stabilize finer particulate matter introduced by currents and provide a protected habitat for a variety of organisms. In lower latitudes *mangrove* trees attach themselves on low-lying tidelands by means of an open network of woody roots, which also has a stabilizing effect on the shoreline.

The rise and fall of the tide shifts the zone of wave activity alternately shoreward and seaward. Areas of swash, surf and breakers are systematically shifted, introducing periodic rhythms to the forces of shoreline deformation.

ABNORMAL EVENTS

The normal work of wind, waves and tides is interrupted from time to time by unusual events of titanic proportions. One of these is the *storm wave*, or better, *storm surge*, an abnormal rise of the sea produced by large, deep cyclonic storms. The very low central atmospheric pressure and the contracting spiral circulation of surface winds force the sea to rise beneath the migrating vortex. When a storm surge reaches land it is all but impossible to distinguish in the general fury of pounding waves, high winds and heavy rains of the storm. Its effects are compounded when the arrival of a storm coincides with high tide at the time of the new or the full moon, especially along a low-lying coast. Exact heights for storm surges are not known, but are estimated to range up to 20 feet or more. In the New England hurricane of 1938, the sea surface rose some 15 feet from the combination of all causes, leaving the shoreline in its wake a scarcely recognizable wreck. At the head of the Bay of Bengal in 1737, the water level rose about 40 feet, sweeping in over the low Ganges delta to cause some 300,000 deaths. Barrier beaches are breached, new inlets created, former channels filled in and a general rearrangement of the shoreline appears. By this means coastal modification is hastened and a new stage of coastal interaction dramatically begins.

Another unusual occurrence of profound importance to coastal interaction among land, sea and air is the *seismic sea wave* or *tsunami* (Japanese: *tsu,* harbor; *nami,* wave(s)). This is a long, low wave set in motion by large submarine disturbances such as earthquakes, volcanic eruptions or landslides. Its length from crest to crest is normally between 80 and 150 miles, its height usually less than three feet, its average period about 16 minutes. It moves with extraordinary speed. This depends on depth, and in the central Pacific, where mean depth is over 15,000 feet, the velocity averages over 470 miles per hour. Theoretically it can attain speeds in excess of 600 miles per hour. In shallow water tsunami move much more slowly, their speed diminishing gradually with decreasing depth. At the same time wave height increases, its actual rise depending on the slope of the bottom, shoreline configuration, and the direction of wave propagation.

On April 1st, 1946, a severe earthquake occurred in the Aleutian Trench. It sent out seismic sea waves in all directions from the *epicenter,* the point at the earth's surface above the earthquake's focal point. Speeding at more than 470 miles per hour at the surface of the open sea they rose to great height on nearby shores, reaching more than 100 feet on Umiak Island, destroying a lighthouse. At one point on the Hawaiian Islands, some 2,000 miles south where the tsunami struck about 4 hours later, the sea rose 57 feet. Coastal damage was extensive here and in Japan, totaling over $10,000,000, and 173 lives were lost. The tsunami of the Alaska earthquake on March 27, 1964, brought severe damage to coastal features as far south as Eureka harbor, in northern California.

From 1928 to 1963, 84 tsunamis were reported throughout the world, of which 66, about 80 percent, were in the Pacific. Five were widely destructive and 44 caused no damage. Most appear to originate in the deep trenches that rim the Pacific Ocean basin as the result of earthquakes in which faulting, the slipping of one mass of bedrock against another, is the generating movement.

Tsunamis have also been created by volcanic explosions. Krakatoa, a small island between Java and Sumatra, exploded in 1883 sending out sea waves that rose to 135 feet. Nearby shores of Sumatra and Java were inundated, receiving heavy damage and tremendous loss of life. The oceanic pulsations from this explosion were recorded on tide range gages around the world, perhaps as far as the English Channel. An estimated 18 cubic miles of rock disappeared from the island and clouds of dust and ash shot up to the stratosphere. The influence of volcanic dust on the qualities of the atmosphere is dealt with more fully in the next chapter.

Major landslides also set tsunamis in motion when they occur along the coast or beneath the sea. Coastal landslides severely altered details of shoreline configuration in southern Chile, following the earthquake of May 22, 1960. The work of deformation was augmented by subsequent wave action. A similar set of forces operated around Anchorage, Alaska, in 1964. An earthquake in southern Alaska in July, 1958, occurring along a fault at the head of Lituya Bay, sent a seismic sea wave down the bay whose crest was about 100 feet high. Large landslides of heavy rock cascaded down the steep sides of the bay into the fiord waters below. Opposite the largest rockslide a gigantic swash had sent a tongue of water 1700 feet up the slope.

Extraordinary events responsible for sudden coastal deformation are mainly of tectonic origin. Earthquakes, landslides and tsunamis tend to displace portions of the land into the sea. Volcanoes, on the other hand, with the exception of those like Krakatoa, normally contribute ash and lava to existing land, thus acting in a constructive way opposing erosive processes. Most volcanic activity occurs around the perimeter of the Pacific, almost entirely encircling the great ocean basin. This distribution is sometimes referred to as the *rim of fire*, and is more or less parallel to the distribution of deep ocean trenches. There are many more volcanoes in the ocean basins than on the continents, and most of them are in the Pacific. Within historic time a great many small islets have been created by the emergence of a volcanic peak above the sea surface. This is the creation of new land, and on a very minute scale, new shorelines; new loci of interaction among land, sea and air. Of those that have appeared in the past two or three decades, the most recent is the island of Surtsey.

On November 14, 1963, at a point about 21 miles southwest of Iceland, a great cloud of smoke, steam and ash billowed up from the sea surface. In one day an ash cone had formed more than 30 feet high; three days later it was about 2000 feet long and nearly 200 feet high. By the following April lava had begun to flow over the ash surface, and by April, 1965, had attained a circular shape with a diameter of about one mile. An ash cone stood about 550 feet above sea level and a composite cone of ash and lava reached upward almost 400 feet. Mosses and lichens have since become attached, and sea birds have begun to roost there. In time, nutrients from salt spray cast up by the waves, and accumulated from the droppings of the birds, along with their feathers, will enrich the organic material to nourish other plant and animal life enabling them to take hold. This result of the interaction of land, sea and air along the new shores of this small island will evolve slowly. A more immediate result was the occurrence of heavy showers during the early days of the island's formation, as steam clouds rose into the chill, damp air south of Iceland and were rapidly cooled to the condensation point.

REFERENCES

BASCOM, W., "Beaches", *Scientific American,* Vol. 203, No. 2, 1960.
BERSTEIN, J., "Tsunamis," *Scientific American,* Vol. 191, No. 2, 1954.
KING, C. A. M., *Beaches and Coasts,* London: Edward Arnold, 1959.
LAUFF, G. H., (Ed.), *Estuaries,* Washington: American Association for the Advancement of Science, 1967.
MACMILLAN, D. H., *Tides,* New York: American Elsevier Publishing Co., 1966.
RUSSELL, R. J., *River Plains and Sea Coasts,* Berkeley: University of California Press, 1967.
TRICKER, R. A. R., *Bores, Breakers, Waves and Waters,* New York: American Elsevier Publishing Co., 1965.

6

Interaction Between Atmosphere and Land

TOPICS

Action of land on air   Significance of air/land interaction
    Action of air on land   in the geosystem

In many ways the nature of the interaction between land and air is distinctly different from that between air and sea. This arises in part from the great contrast between the sea's mobility on the one hand and the resistance of the solid land to movement on the other. The mobile fluidity of the sea is immeasurably more consonant with the acutely sensitive mobility of the atmosphere. On land where topography directs the flow of surface airstreams even the most powerful winds conform to the configuration of the land. Furthermore, the sea presents a homogeneous surface to the atmosphere, while land terrain is exceedingly varied in quality, slope and altitude.

The endless array of surface material is exposed to the atmosphere at countless angles and in countless attitudes within elevations ranging from 1,286 feet below sea level (the Dead Sea) to 29,028 feet above sea level (Mt. Everest). An incalculable variety of plant associations and soils compounds the complexity of terrestrial terrain. In addition, all are continuously subjected to processes of change. The primary significance of land surface variations in the geosystem is their geographical distribution. This will become evident as the discussion of land/air interaction unfolds in this and the following chapter.

It is expedient to discuss the exchange of energy, matter and momentum between air and land in three steps: 1) the action of land on air; 2) the action of air on land; and 3) significance of air/land interaction in the geosystem.

THE ACTION OF LAND ON AIR

Energy transferred from land to air is almost entirely that of sensible and latent heat initially provided at the surface by global radiation as discussed in Chapter 2. It is released to the atmosphere by conduction, convection, radiation and evaporation. Evaporation far exceeds all other energy transfer processes from lakes, streams, marshes, and the foliage of living plants. It is also the chief transfer process from all solid surfaces that are moist. Thus in humid regions where precipitation is greater than potential evapotranspiration, evaporation is the main means of energy exchange from land to air. In dry regions, however, where evaporative power exceeds the available surface moisture, the main energy exchange process is the radiation of sensible heat. Deserts, therefore, are the main regions of radiative heat exchange. And when in humid regions the surface of the land becomes dry, radiative heat transfer becomes the leading process. Data from O'Neill, Nebraska, illustrate this important point.[1] Table 6.1 shows the disposition of incoming radiation for two days in the summer of 1953, on one of which the soil surface was moist and on the other dry. August 9th was preceded by rain, the soil surface was moist and evaporation accounted for 65 percent of the heat transfer to the atmosphere (29 percent plus 54 percent). September 7th was preceded by a dry spell, the soil surface was dry and evaporation accounted for only about 21 percent of the heat exchange to the atmosphere (19 percent plus 70 percent), while sensible heat transfer by radiation accounted for 79 percent. The site was a tract of level grassland, the sky clear with light wind from the south on both dates. The dry soil surface attained higher temperatures on September 7th, even though the net radiation available for heating the soil was significantly less. And on that date air temperature at 4 inches above the surface rose about 40 degrees from the early morning minimum, while in the more humid situation on August 9th the temperature rise was only 31 degrees.

Evaporation is the transfer of both matter and latent energy. It is estimated that about 70 percent of the precipitation falling on the world's land areas is returned to air by evaporation. The surfaces from which water vapor is released are infinitely varied and their physical condition is continually changing. Thus, the exact part played by each surface cannot be accurately assessed, and only approximations are available. As an example, it has been calculated for the German Federal Republic that the year's total precipitation of 771 millimeters in 1951 was disposed of as follows: 38 percent was evaporated from vegetation, 13 percent evaporated from the ground, 39 percent entered streams and the remainder went into the ground water. At Eberswalde, northeast of Berlin, annual evaporation from bare ground, short grass, young pines and meadow underlain by a high water table, is in the ratios of 2:4:5:8.

---

1. U. S. Air Force, Air Research and Development Command, *Handbook of Geophysics*, revised ed., (New York: The Macmillan Co., 1961) pp. 2-4; 2-17.

TABLE 6.1

*Net Radiation and Heat Budget on Level Grassland
at O'Neill, Nebraska, on August 9 and September 7, 1953*

August 9	04	06	08	10	12	14	16	18	20	Daily Total	Per cent
Net radiation (Watts/m²)	−59	47	364	497	540	525	273	−13	.... ..	2174	
Heat flux into soil	−40	29	186	63	74	73	28	−65	.... ..	348	17
Heat flux into air	−11	....	81	158	176	190	64	−17	.... ..	641	29
Heat equivalent of evap.	− 8	....	97	276	290	262	181	69	.... ..	1167	54
Air temperature (°F) at 4″	58.5	62.2	73.4	81.9	87.3	89.0	89.6	84.0	76.1		
Soil temperature (°F) at 0.2″	63.6	64.2	73.4	88.0	96.0	97.2	92.9	84.7	77.9		

September 7	04	06	08	10	12	14	16	18	20	Daily Total	Per cent
Net radiation	−54	−32	181	403	488	398	154	−69	−77	1392	
Heat flux into soil	−64	−25	36	84	95	66	13	−29	−28	148	11
Heat flux into air	− 6	− 6	98	230	303	299	114	−30	−39	963	70
Heat equivalent of evap.	− 4	− 1	47	89	90	33	27	−10	−10	261	19
Air temperature (°F) at 4″	46.6	49.1	65.8	76.8	83.7	86.0	82.6	73.6	66.8		
Soil temperature (°F) at 0.2″	53.3	52.9	65.1	86.2	100.0	99.1	88.4	76.7	69.4		

From U. S. Air Force, Air Research and Development Command, *Handbook of Geophysics*, rev. ed. (New York: The Macmillan Co., 1961) pp. 2-4; 2-17.

This shows that evapotranspiration from moist meadow yields latent energy at a rate four times greater than that of bare ground.[2]

The amount and rate of direct (sensible) heat transfer from land to air depends on the amount and rate at which incoming radiation is concentrated at the surface. This in turn depends on the qualities of the ground surface materials and includes color, density, porosity and subsurface moisture content, as well as organic content and the character of plant cover. These qualities essentially determine the heat absorptivity, albedo and conductivity of the substance. This is illustrated

2. Geiger, R., *op. cit.*, p. 278.

in Table 6.1, showing the strength of radiative heat flux from dry soil surface to air in the magnitude of air temperature change between 4 a.m. and 2 p.m. on September 7th. Table 2.4 in Chapter 2 gives the absorption of incoming radiation by various surfaces in loosely generalized percentages for purposes of comparison.

The importance of the ground surface as a radiative heat source is indicated by values obtained at widely separated points. In southern Finland around the middle of August, the surface of a sandy heath on a sunny day warmed from 41°F at dawn to 93°F by early afternoon, and on a later date rose to 145°F. At the University of Leipzig in East Germany on a clear day in July the surface of natural ground warmed from 58°F at 0300 to 104°F at 1500. Near Palermo, on the northwest coast of Sicily, over 1500 miles south of Helsinki, Finland, at a point 1,394 feet above sea level, surface and air temperatures were measured on a clear day in July. Soil surface values rose from a minimum of 64°F at around 0400 to 144°F at 1345, a gain of 80 degrees. Air temperatures at 4 inches above the dry plant litter rose from 64° to 117° in the same time interval.

Surface heat energy concentration is also profoundly affected by the direction and steepness of slope as well as the elevation of the surface. For example, in the Otzal, of the Austrian Alps, surface temperatures were measured in dark, raw humus without plant cover on a southwest slope with an angle of 35 degrees, during the hot July of 1957.[3] The value was 176°F. On the notheast slope, however, the surface temperature was only 73°F. The altitude was almost 6,800 feet. At this elevation the rarified atmosphere is much less absorptive than at sea level, and the air temperature about 6.5 feet above the highly heated dark humus surface was 86°, less than half that on the ground beneath.

Solid matter in the atmosphere originating at the earth's surface is largely taken up from the surface by advection and eddy diffusion. The solid substances comprising atmospheric dust, as mentioned in Chapter 2, include smoke, gases, pollen, spores, bacteria and mineral particles. But forces within the lithosphere, operating independently of atmospheric motion, eject moisture, gases and particulate matter from the solid earth into the atmosphere. Hot springs and volcanoes are the main sources.

Nearly 500 volcanoes are more or less continually active throughout the world, contributing smoke, ash, lava and steam to the atmosphere at varying rates. The finer effusion components are believed held in suspension indefinitely. Composition of released material differs widely from one volcano to another, but usually includes minute metallic particles,

3. Geiger, R., *op. cit.*, p. 447.

ammonia compounds and sulfur compounds, among them sulfuric acid droplets.

Historically, volcanic eruptions have seriously altered the composition of the atmosphere and induced significant changes in weather. Higher wind speeds near the tropopause are capable of keeping in suspension the finer particulate matter indefinitely, and spreading it widely around the earth. In 1814 and 1815 Tambora Volcano on the island of Sumbawa, Indonesia, erupted bringing total darkness within a radius of over 300 miles for three days, and deepening the intensity of twilight over many parts of the world for some time afterward. The eruptions on Krakatoa in 1883, and at Mt. Katmai in 1912, produced similar results. In 1963, Agung Volcano on Bali, Indonesia, ejected huge volumes of material into the atmosphere that were observed as twilight clouds at 12 to 51 miles over Colorado, in the lower atmosphere. This is somewhat above the *Junge layer*, at 9 to 12 miles in altitude, where stratospheric dust persists continually. Residence time of dust in the stratosphere is believed to be from 5 to 10 years, and apparently volcanic dust tends to settle in the Junge layer. The effect, following severe explosions, is to diminish incoming solar radiation by about 5 percent.

The momentum of air is greatly modified by the roughness of land surfaces. This is the *friction effect,* and is estimated to consume nearly one percent (0.7) of the atmosphere's energy. For the world as a whole the *friction layer* is generally assumed to be the first 3,000 feet or so of the atmosphere above a surface. At sea wind-driven waves exert a frictional drag on the very winds that create them. But the degree of roughness on land which increases in proportion to the magnitude of surface relief, is many times greater than that produced by the surface of the sea. Level, grass-covered plans are scarcely more disruptive to smooth airflow than ice or water surfaces, but the roughness effect becomes greater in shrub-covered plains, and in forested lowland it is still greater. Thus vegetation alone induces kinetic energy loss in moving air. Geomorphic features, the limitless variety of diversified topography, exert the strongest effects, however.

In a general way, wind approaching an elongated mountain range at right angles is slowed by upward deflection, but gains speed and turbulence on the downslope in the lee. Mountain passes constrict airflow, thereby increasing its velocity in response to the law of conservation of momentum. Nocturnal cooling sets in motion downslope winds in all areas of uneven terrain. In hilly country in the middle latitudes the downward settling of cold air at night produces *frost hollows* or front pockets where cold air is impounded. In areas of strong relief and steep slopes, cold air not infrequently moves downslope with gale force creating what is sometimes termed an *air avalanche.* These are *katabatic winds*, gaining kinetic energy as they sweep downward to lower levels. Off the icy plateaus of Greenland and Antarctica, katabatic winds are known to travel down the steep, seaward margins at speeds in excess of 100 miles per hour.

THE ACTION OF AIR ON LAND

Precipitation is the most important effect the atmosphere exerts upon the land. The consequences of this action in the operations of the geosystem are of such significance and such complexity that they must be treated at length in the chapter following this one. Falling precipitation in all its forms simultaneously transmits energy (kinetic and potential), matter and momentum to the surface. It is the principal form of matter transferred from air to land. The momentum of moisture particles, either liquid or solid, plays a minor, but significant, role in erosion of solid surfaces by impact, through the application of kinetic and potential energy.

Heat energy transmitted from air to land is chiefly sensible heat through long wave reradiation both day and night. It is especially effective at night when it offsets the nocturnal radiative heat loss from land surfaces to outer space, limiting the degree to which surface temperatures decrease. The atmosphere also performs the important function of transporting warmth by advection. This is particularly characteristic of the middle and upper latitudes and occurs at all seasons of the year. The effect is most pronounced, however, with the change of seasons from winter to summer, when poleward surges of warm, humid air usher in the transitional season of spring. On such occasions snow may be melted rapidly, ice begins to disappear from streams and ponds, and the land often steams with the release of surface moisture.

A further heating effect occurs in areas of strong relief where prevailing airflow sweeps downward after rising up the windward slopes of high mountain ranges (Figure 6.1). Having reached condensation levels and released much moisture by precipitation, the now drier air moves down the lee slopes and gains heat adiabatically. Increasing its evaporative power as it becomes warmer, it normally exerts a warming, drying influence on the lower surfaces as it expands downward away from higher terrain. Moisture is taken up readily and when the ground is snow-covered much melting occurs as temperatures rise rapidly. Such downslope warming is known as the *föhn effect*. In the Rocky Mountains föhn winds are called the *chinook*. In places within less than 50 miles of the eastern ranges the thermometer may rise 40 degrees in three hours, and in Alberta the temperature has risen 40 degrees in 15 minutes. Snow may be melted away at a rate of nearly one inch per hour.

Figure 6.1. The föhn effect. Warm, humid air, rising up the windward mountain slopes, cools at the wet adiabatic rate of, say, 3.3°F per thousand feet. Descending the lee slopes, it warms more rapidly at the dry adiabatic rate of 5.5°F per thousand feet.

Another form of energy transmitted from air to land is lightning, which occurs during mature thunderstorms when the normal electrical field between earth and atmosphere is reversed. Enormous energies are released, mainly as light and heat, and locally heat from a lightning strike may initiate combustion. About 50,000 thunderstorms are estimated to occur daily throughout the world (see Figure 7.5), and about 2,000 are thought to be in progress at any one time. This suggests a continuous lightning frequency of about 100 strokes per second for the world as a whole, and constant energy exchange of high intensity.

The kinetic energy of atmospheric momentum is sufficient to lift enormous volumes of solid material from the face of the earth. The most powerful applications of this energy are developed in the tornado which has the capacity to lift tons of material suddenly and rapidly from the surface. Tornado wind speeds are not known but are estimated to exceed 300 miles per hour and perhaps to reach 500 mph. Updrafts probably reach 100 mph. The resultant wind pressure exceeds 700#/ft².

A barometric pressure of 25 millibars may occur within the center of the swiftly swirling vortex. The pressure drop plus extraordinary wind velocities destroy structures, uproot trees, and raise dust and debris high into the air, dispersing it far from the point of origin. Over 90 percent of all recorded tornadoes occur in the United States, most of them in the plains region east of the Rocky Mountains, although they are not uncommon in parts of Australia and Africa. A companion phenomenon is the water spout which forms most commonly over warm, tropical seas, and is capable of raising water up to ten feet above the surface, and spray several hundred feet into the air. Tropical cyclones (see Figure 7.6) also expend their energies on portions of the land surface, although soon after leaving the sea they tend to lose vigor, usually dissipate rapidly and die out. The main effect of a mature hurricane on land is the flooding that follows unusually intense rainfall. Tornadoes may develop around the perimeter of a large hurricane, compounding the effect of its momentum on the surface of the land.

The effect of atmospheric momentum on land surfaces is generally much less rapid and much less dramatic than is suggested by the high winds of violent storms. Its capacity to keep very fine dust particles in suspension has been discussed earlier. The dimensions of such particles are measured in microns. Much of the atmosphere's constant load of suspended sediment is cast into the air by volcanic effusion, by the spray of breaking waves, and as the smoke of combustion. But the atmosphere also has the capacity to raise particulate matter from the earth's surface. Dry land is thus the chief source of atmospheric dust.

The distance from the point of removal at which particulate matter falls back depends on particle size, shape and density, as well as wind speed, duration, steadiness and fetch, but principally wind speed. Airborne particles are usually no more than 2 millimeters in diameter and are classed by geologists as clay, silt or sand, according to size. Clay is less than 4 microns, silt is from 4 to 63 microns, and sand ranges from 63 microns to 2 millimeters. Wind speed of 5 meters per second (slightly more than 11 miles per hour) is the minimum required to move sand along the ground. Horizontal winds of that speed produce upward eddies of one-fifth that speed (1 m/sec), and this is just enough to keep in suspension particles smaller than coarse silt (less than 63 microns). (See Table 6.2.)

Wind is potentially effective on all surfaces of loose, dry material where vegetation is lacking, including desert regions, sand beaches, and plowed farmland in time of drought. By removal, transport and deposition wind is capable of significantly modifying the surface. Removal (called *deflation*) may create *deflation basins*, as in the semi-arid grasslands of central North America from Alberta to Mexico. Although in humid periods the grasslands are protected from erosion by a mantle of short grass, they become vulnerable to wind action in time of drought. The basins are usually only a few feet deep and less than one mile long, but they can be deepened to 150 feet, as in southern Wyoming, depending on the depth of the water table, the upper limit of continuously wet soil.

Deflation attained titanic proportions on the Great Plains during the exceptionally dry period of the 1930's. In places the soil surface was

TABLE 6.2

Particle size of sand, silt and clay

Sand	Millimeters		
Very coarse	2	to	1
Coarse	1	to	.5
Medium	.5	to	.25
Fine	.25	to	.125
Very fine	.125	to	.0625

Silt	Microns		
Coarse	63	to	31
Medium	31	to	16
Fine	16	to	8
Very fine	8	to	4

Clay	Microns		
Coarse	4	to	2
Medium	2	to	1
Fine	1	to	.5
Very fine	.5	to	.25

lowered about 3 feet in as many years, partly because of soil moisture deficiency and partly from the abnormally high frequency of strong winds. At times dense dust clouds extended from ground level to 12,000 feet producing almost total darkness as they passed. Air-borne solid material, estimated on one occasion to equal 166,000 tons per cubic mile of air, was widely distributed. Unusually deep haze filled the sky to produce a twilight effect as far east as New England, some 2,000 miles from the source regions along the eastern front of the Rockies. In Nebraska deposition is calculated to have been on the order of 800 tons of solid substance per square mile.

Deflation, transport and deposition are most extensive in desert regions. Most desert floors are rocky, and the removal of finer materials gradually leaves a pavement of rock fragments too large for the wind to dislodge. But where sand and finer particles accumulate they become drifted into rippled dunes. These are then subjected to frequent modification, readily responding to the frictional drag of the wind and the transport power of its momentum. A similar result of atmospheric motion along the air/land interface is the drifting of dry snow in winter.

Wind-driven particles exert a gradual erosive effect on prominently exposed rock outcrops often producing smoothly sculptured features in the world's dry regions. And when fine, dry material, picked up from the desert floor, is deposited in more humid, peripheral areas by the prevailing winds for periods measured in thousands of years, it often accumulates in depths of 100 feet or more. These deposits of uniformly fine texture, usually free of larger rock fragments, become closely compacted and firm, are light in color and are known by the term *loess*.

SIGNIFICANCE OF AIR/LAND INTERACTION IN THE GEOSYSTEM

Interaction between air and land is one phase in the dynamics of the geosystem. The exchange of heat, moisture and dust particles is the most important direct interaction at the interface. The intensity of the exchange varies within very wide limits of both time and place. The variance of exchange intensity is attributable to the widely ranging size, shape and distribution of land surfaces, plus the diversity of surface configuration. Thus the geography of exchange intensities profoundly influences the character and magnitude of air/land interactions. Those interactions are both a product of geosystematic operations and an influence on them. The geosystem is a complex of feed-back mechanisms dynamically integrated.

The solid surface of land rapidly concentrates the heat of global radiation when its intensity increases and loses it rapidly when it diminishes. This capacity is the basis of the great contrasts arising between land and sea in the functions of the geosystem. The atmosphere overlying the land is accordingly treated to a sequence of alternate heat surplus and heat deficit. This is one of the chief contributions land surfaces make to the functions of the geosystem. Alternate heating and cooling take place on both a diurnal and a seasonal basis.

Diurnal heating and cooling proceeds most effectively when the air is clear and dry. Its effectiveness is a conspicuous feature of stable, sunny weather in the mid-latitudes during summer and in the tropics throughout the year. In the tropics visible evidence of diurnal heating commonly appears through the deepening of cumulus clouds when the conditionally stable air of the trade winds passes over land (Figure 6.2). This is a striking feature of even very small, low-lying islets such as the coral reef atolls. The location of small islands beyond the horizon is often revealed by the presence of cumuli towering well above the scattered ranks of the prevailing trade wind cloud deck over the surrounding sea. The clouds are the product of thermal convection circulations that begin soon after sunrise, often deepening noticeably by mid-day. Trailing off to leeward in more or less well-defined linear distribution, they are known on occasion to produce rain. The effects of such rain-producing cumulus clouds were observed as they were generated over the island of Anegada on March 26, 1953.[4]

Anegada, a small, low islet in the West Indies about 80 miles northeast of Puerto Rico at a latitude of 18° 45'N, is fully exposed to the steady flow of the trades. About 10 miles long and 2 miles wide, its maximum elevation is around 30 feet. On March 26th, oceanic cumulus were sparse and stunted, rising from a 1500 foot base to less than 2000 feet. But the island clouds, observed from around noon to the later afternoon, formed a single, long street, parallel to the trade flow. Beginning at the windward end they extended for perhaps 15 nautical miles beyond the leeward end of the island (Figure 6.3). They were much taller, reaching from a higher base of about 1800 feet to more than 7500 feet. One of them exceeded 8000 feet. Downwind from the island rain fell for about two hours during the early afternoon from three or four dense cumulo-nimbus clouds spaced about 3 nautical miles apart. At sundown normal trade wind clouds replaced the island cloud street following withdrawal of the generating heat transfer at the air/land interface.

The sea breeze is another product of diurnal interaction between air and land. Again it occurs most effectively under clear, comparatively stable atmospheric conditions. The sea breeze is a common feature of coastal weather in the tropics the year 'round, and in the mid-latitudes during summer. Regularity and intensity vary widely, depending on the character of prevailing larger scale atmospheric states, and the configuration and exposure of the shoreline. The sea breeze normally consists of a distinct onshore flow during the day when land becomes much warmer than the surface of the adjacent sea, and a less vigorous, ill-defined offshore drift at night when the land becomes cooler than the sea. It begins soon after sunrise, strengthening progressively as the land warms rapidly to the increasing intensity of global radiation. Diurnal onshore fluxes of maritime air usually reach inland no more than 10 to 15 miles, although sometimes they penetrate more than 50 miles. They are a recognizable phenomenon as far north as the Baltic Sea. Onshore movement is typically confined to a lower stratum less than 2000 feet deep, and a weaker, compensating seaward flow develops above it.

---

4 Malkus, J. S., "Tropical Rain Induced by a Small Natural Heat Source," *Jour. Appl. Met.*, Vol. 2, No. 5, Oct., 1963, pp. 547-556.

Figure 6.2. Trade wind air, flowing westward (from right to left) over the Florida peninsula is relatively stable approaching the Atlantic coast. Passing over the land it is buoyed up by intense heating of the surface below, reaching condensation levels between 2,000 and 3,000 feet to produce long lines of cumulus clouds many of which become rain-producing cumulonimbus. Passing offshore over the Gulf of Mexico, the air is cooled from below, loses buoyancy and cloud formations disappear. (NASA photo).

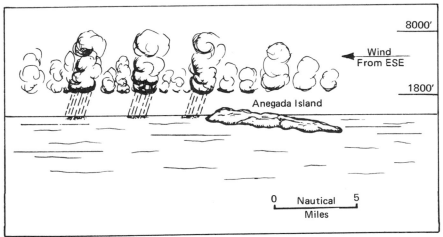

Figure 6.3. Anegada, low islet about 80 miles northeast of Puerto Rico. Rain-producing cumulonimbus clouds formed by thermal convection resulting from intense heating of moist, oceanic air passing over land on March 26, 1953.

The sea breeze is singularly characteristic of summer weather along the shores of the Mediterranean Sea. Its effectiveness, dimensions and schedule vary considerably from place to place. However, the process is somewhat as follows: from sunrise to 0900, calm; 0900, sea breeze begins; 1400, sea breeze maximum; 1800, sea breeze ends; 1800-2000, calm prevails; 2000, offshore breeze begins; around sunrise, offshore land breeze ends. The layer of onshore flow is normally about 1500 feet deep, but occasionally reaches around 3000 feet and may travel up to 50 miles inland. The offshore nocturnal drift is seldom more than 500 feet deep and its effect is usually felt only a few miles at sea. The strength of onshore flow is normally several times greater than the off-shore drift, attaining speeds of between 10 and 25 miles per hour. Beginning nearly perpendicular to the shoreline, onshore flow tends to veer around to approach at an angle of about 45 degrees when maximum speed is attained in the early afternoon.

In general the sea breeze provides tangible evidence of its occur-rence by providing moisture necessary for thermal convection to produce cumulus clouds. Cloud cover thus increases a few miles inland while skies remain clear near the water's edge. At the same time a refreshing temperature drop occurs. This is especially evident along many mid-latitude coasts in early summer. Along the north shore of Long Island Sound the temperature drop averages about ten degrees with the onset of the sea breeze. But from time to time the change is twice as great. At Noank, Connecticut, on May 12, 1958, for example, the thermometer showed a drop of 20 degrees from 72°F to 52°F in less than five minutes in mid-afternoon with an abrupt shift to an onshore wind.

On a seasonal basis the monsoon is exactly analogous to the diurnal sea breeze. In this the interaction between atmosphere and land pro-duces a periodic airflow reversal of much larger scale. A marked seasonal shift in prevailing surface wind direction takes place in many parts of the world. But the best known example of broad scale directional reversal is the Indian Ocean monsoon between southern Asia and north-ern Australia. The sinusoidal reversal of direction between July and January results primarily from the alternate heating and cooling of the atmosphere over southern Asia and Australia. Onshore pulsations of humid, maritime air bring a period of copious rains, while offshore movement during the cool season brings an interval of drought. Mon-soonal response to the influence of alternate heating and cooling on land is also well-developed along the Guinea Coast of west Africa, and to a lesser degree along the Pacific coast of Mexico and Central America, and the Gulf coast of the United States.

On an even larger continental scale the seasonal rhythm of air/land interaction is plainly evident in the maps of temperature distribution for July and January (Figure 3.4). In the northern hemisphere, for example, as summer advances, land surfaces respond rapidly to the increasing intensity of incoming radiation, becoming much warmer by July than oceanic surfaces at the same latitudes. Thermally uplifted air over the continents brings a consequent lowering of atmospheric pressure as air aloft tends to move off to seaward. Cooling as it leaves the source of its excessive heat supply, it becomes denser over the sea, subsides and in part augments the densities of the semi-permanent oceanic highs. These in turn intensify and expand over the continental margins. Their poleward flow transports successive surges of humid, maritime tropical air well into the interior of eastern North America and eastern Asia. From this highly simplified view, large land masses are seen as surplus heat sources in summertime, significantly affecting the surface circulation.

Maximum summertime temperatures are attained in the arid regions of the lower middle latitudes. Here atmospheric temperatures near ground level are little affected by influxes of maritime air and repeatedly exceed 100°F. At El Azizia, Libya, the world's highest air temperature of 136°F has been recorded, and in Death Valley the maximum has reached 134°F. At Marble Bar, Western Australia, the mean maximum temperature was 110°F for 106 consecutive days.

In winter the function of land/air interaction in the geosystem is reversed. Land areas in the higher middle latitudes of the northern hemisphere become source regions of dry, clear, deeply cold air or *heat sinks*. Heat loss begins to exceed heat gain after the autumnal equinox. During the coldest month of January the interior of northern North America normally experiences mean temperatures lower than −20°F, and at Snag, Yukon, the termometer has dropped to −81°F. In eastern Siberia, mean January temperatures are less than −50°F. At Verkhoyansk the normal value is −58°F and an extreme minimum of −94°F has been observed.

South of the equator seasonal thermal fluctuations are much less pronounced due to the smaller land areas within the middle latitudes. Thermally conservative maritime air dominates nearly all of the southern hemisphere. Nonetheless even here the mean annual temperature range is considerably greater over land than over the sea within the same latitudes.

Annual thermal variations over major land areas strongly influence the direction and intensity of seasonal shifts in atmospheric circulation (see Figure 3.5). These in turn profoundly affect the seasonal distribution of the world's precipitation. The varied distribution of heat intensities, the effect these have on atmospheric circulation, and the resultant distribution of precipitation is a dynamic integration of primary significance in the geosystem.

A unique function of air/land interaction in the geosystem is the direct transport of dust from land to sea by wind. Most atmospheric dust deposited at sea consists of minute mineral particles derived from dry, exposed soils and the persistently dry surfaces of the world's deserts. However, pollen, spores and bacteria are organic forms that are carried far from shore, are deposited on the surface, and add to the plankton and the chemistry of the sea. Borne well up into the troposphere by eddy diffusion organic particles may remain indefinitely in suspension, especially the spores of certain mosses and the puffball. Volcanic effusions continually supply the atmosphere with dust and when great eruptions occur, as discussed previously, they become major sources of fine, wind-borne material transported to sea. Most material ejected from volcanoes is deposited within a few miles of the source. Once deposited on the sea surface, particulate matter may be transported far before settling to the bottom. Much material is dissolved before reaching the ocean floor. It is estimated that quartz (silicon dioxide) particles about 10 microns in diameter require two years to reach bottom at around 13,000 feet. The Pacific is a region of much more volcanic activity than the Atlantic. Bottom deposits in the central Pacific contain a high percentage of volcanic components that are extremely rare on the floor of the Atlantic.

It is easily understandable that when exposed soils are dry, even in normally humid regions, they readily yield variable amounts of dust to the atmosphere, depending in part on wind speed and turbulence. Theirs is an important contribution at all times and is greatly increased in times of severe drought. But the main source regions of atmospheric dust are the world's deserts. The function of wind-transported dust from continental deserts is a significant phase in the operations of the geosystem. The Sahara is the largest continuous tract of desert contributing mineralogical dust in the interaction between air and land.

The Sahara is a virtually unbroken desert tract extending nearly 3500 miles eastward from the Atlantic to the Red Sea, and some 1000 miles from the Mediterranean coast of Libya southward to the basin of Lake Chad. Dust from the Sahara is carried northward across the Mediterranean into southern Europe, usually in powerful surges of southerly flow ahead of migrating cyclones moving eastward. This airflow normally develops maximum intensity in spring, and while it bears many local names, it is best known as the *sirocco*, and is often identified by *mud rains*, or *blood rains* that color the snows of the south-facing mountain slopes. Sahara dust also falls into the Mediterranean, adding to the salinity of the sea. Mediterranean water chiefly gains salinity by excessive evaporation during the dry summer season, and drifts into the Atlantic through the Straits of Gibraltar a few hundred feet below the eastward surface flow from the Atlantic. It contributes its higher temperatures and salinities to the global oceanic circulation as it sinks to depths of around 3,000 feet or more.

During the northern winter the *harmattan,* a frequent strong, warm wind off the desert, carries Sahara dust southward over the Gulf of Guinea, at times creating an oceanic haze so dense as to impede navigation. At this season, desert dust carried out to sea by the harmattan is taken up in the southeast trades across the Atlantic to the northeast coast of South America.

The seaward spread of dust off the western Sahara has long been known among mariners sailing the waters between the Canary Islands and Cape Verde. A deep, reddish haze is a frequent occurrence and rains have often reddened the decks, sails and rigging of sailing vessels over the past several hundred years. Recent observations at Barbados, easternmost island of the West Indies, indicate that dust originating in the Sahara is borne westward in the steady flow of the trades to the islands of the Caribbean.

REFERENCES

BECKINSALE, R. P., *Land Air and Ocean,* 3rd ed., London: G. Duckworth and Co., Ltd., 1960.
HORROCKS, N. K., *Physical Geography and Climatology,* London: Longmans, Green and Co., 1953.
KENDREW, W. G., *Climates of the Continents,* Oxford: Clarendon Press, 1961.
RUMNEY, G. R., *Climatology and the World's Climates,* New York: The Macmillan Co., 1968.
THORARINSSON, S., *Surtsey* New York: The Viking Press, 1967.

Water Transport
From Land to Sea

TOPICS

Precipitation in the geosystem
Geography of precipitation types

Stream transport
Drainage to the sea

The phases through which energy, matter and momentum are continually exchanged in the geosystem reach a climax in the return of water from land to sea. Flowing streams of water and ponderous masses of ice transport to successively lower levels not only water originally derived from the sea, but also the substance of the land itself. Nourished by the precipitation of water vapor in the atmosphere, streams and glaciers move downward toward the sea, eroding, transporting and depositing the solid material of the very surface that supports their existence. With variable velocity and volume they work, in the long range view, toward the levelling of the land, responding to the force of gravity, and theoretically lowering the surface toward a terminal plain of equilibrium. The rate of denudation in the United States, for example, is from one to three inches every one thousand years.

By delivering water to the sea, streams and glaciers close the cycle of energy and moisture exchange that the circulation of the atmosphere preserves between sea and land. At the same time their mineral burden is released as they lose momentum upon entering the sea. In this way rock material is returned that was laid down during earlier geological ages by former inundations of the sea. Sand that is molded by wind-driven waves and currents into beaches and other shoreline features is mainly supplied by entering streams from the land. Finer suspended particles find their way farther seaward, to settle on the ocean floor. Dissolved materials contribute to the salinity of the sea.

PRECIPITATION

Most of the world's normal annual precipitation falls at sea. Estimates of the actual percentage vary, but range between 76 percent and 80 percent. More precise values on the earth's water balance are anticipated after completion of the International Hydrological Decade, 1965 to 1974. Between 20 and 24 percent of global yearly precipitation reaches land surfaces. Of this, nearly 70 percent is returned to the atmosphere by evapotranspiration. Thus only about 30 percent is available for the work of water transport from land to sea. Nearly all the return of energy, matter and momentum to the sea is accomplished by the flow of rivers and streams. Only a negligible amount results from ground water seepage into the sea and the slow, plastic movement of glaciers extending tongues of ice into the coastal waters of the sea.

The water balance of individual continents is shown in Table 7.1. Approximate values of precipitation and runoff are given in inches, along with the percent of mean yearly precipitation available for stream transport.

TABLE 7.1
Continental Water Balance
(Inches)

	Precipitation	Runoff	Percent
Europe	23 ″	9 ″	39 %
North America	26	10	38
South America	53	19	36
Asia	24	8	33
Africa	26	6	23
Australia	18	2.4	13
Mean	28.3″	9.1″	30.3%

The high percentage of precipitation entering the stream systems of Europe and North America is partly due to the large proportion of those continents within the higher latitudes, resulting in relatively low evaporation. The high value for South America results from the small proportion of desert and the very large expanses receiving heavy rains, notably the Amazon Basin. Australia is affected least by stream runoff since more than half its area is desert and the remainder receives relatively little precipitation.

A somewhat closer estimate has been made of the mean annual moisture budget of the United States, based upon stream flow data from more than 7,000 gaging stations between Canada and Mexico. Mean yearly precipitation for the 48 states has been calculated at 30 inches. Humid, occanic air brings water vapor over the land equivalent to 150 inches of precipitation if it were all released to the surface. Only 30 inches are precipitated, however, and of these, about 21 are returned to the atmosphere by evapotranspiration. The net gain for stream drainage to the sea is about nine inches, or 30 percent. But the proportion of runoff to precipitation varies greatly from one stream system to the next. In the drainage basin of the Ohio, for example, 30 percent of the normal yearly precipitation is runoff, compared with less than 8½ percent for the Colorado. Similar contrasts are typical of every land mass. The primary basis of such contrasts is the uneven distribution of precipitation. The wide range of mean annual rainfall is shown in the map of world distribution, Figure 7.1, averaging over 80 inches in many tropical regions and less than 10 inches over vast expanses of desert. A number of small, isolated areas have recorded average yearly rainfall of over 300 inches and a number of arid localities have yet to record any measurable rain.

But the amount of precipitation received in a given year seldom equals exactly the long term average, and account must therefore be taken of the probable departure from normal. This is shown for the world as a whole in Figure 7.2, where variability is seen to increase as mean annual rainfall decreases. The effect of yearly rainfall fluctuations on stream flow is much more severe than the range of values on the map suggests. For example, the calculated mean precipitation for the Ohio Basin is 42.2 inches. Average moisture return to the atmosphere by evapotranspiration is over 60 percent which amounts to 25.4 inches. This allows a normal yearly supply of 17 inches for runoff. Total precipitation for a given year may easily be 15 percent less than normal. It is often quite a lot less than this. Fifteen percent less than normal precipitation means that 6.36 fewer inches fall in the Ohio Basin. Such a decrease is usually accompanied by an increase in evapotranspiration. But for the purpose of providing a conservative illustration, a normal evapotranspiration loss of 25.4 inches may be assumed. With 36.0 inches falling on the land (42.4 − 6.36), 25.4 inches are released to the atmosphere, leaving only 10.6 inches for runoff. This is a decrease of more than 37 percent resulting from a 15 percent lower input of precipitation.

Seasonal distribution introduces a further variation of great potential significance to stream flow. When abundant precipitation falls more or less uniformly throughout the year, ample water is available for runoff and stream levels fluctuate very little. This is the situation in southeastern Canada where mean annual precipitation ranges between 30 and 50 inches and is comparatively uniform from season to season (see climatic chart for Montreal in Figure 7.3). In consequence, the rivers between the St. Lawrence and Hudson Bay are known for their relatively steady flow. But where marked seasonal changes occur, the amount of water supplied to streams may vary widely. This is a notable feature of many major tropical river systems where most of the year's rain may

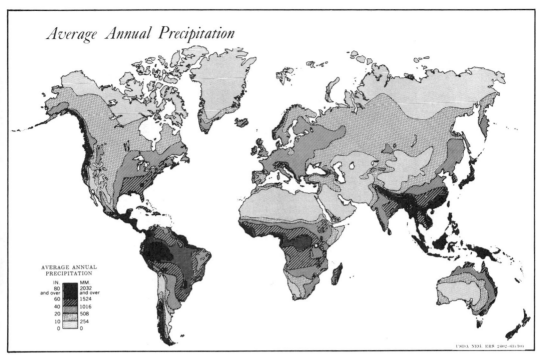

Average Annual Precipitation

AVERAGE ANNUAL
PRECIPITATION

IN.	MM.
80 and over	2032 and over
60	1524
40	1016
20	508
10	254
0	0

USDA NEG. ERS 2402-63(10)

Figure 7.1. Geographical distribution of mean annual precipitation. (U. S. Department of Agriculture map).

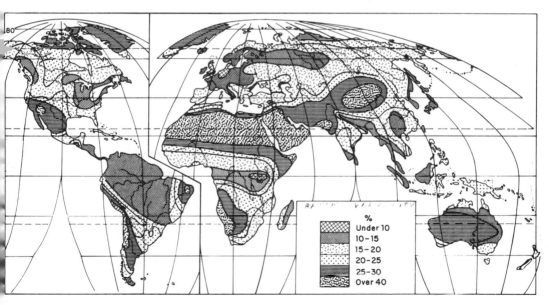

%
Under 10
10–15
15–20
20–25
25–30
Over 40

Figure 7.2. World distribution of normal annual rainfall variability. (After Rumney, G. R., **Climatology and the World's Climates,** New York, The Macmillan Co., 1968, p. 396).

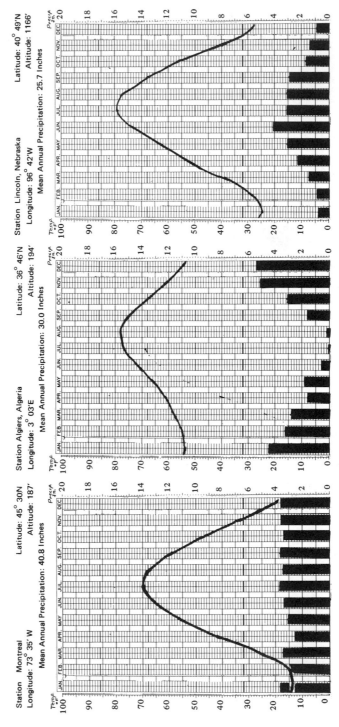

Station Montreal   Latitude: 45° 30'N
Longitude: 73° 35' W   Altitude: 187'
Mean Annual Precipitation: 40.8 Inches

Station Algiers, Algeria   Latitude: 36° 46'N
Longitude: 3° 03'E   Altitude: 194'
Mean Annual Precipitation: 30.0 Inches

Station Lincoln, Nebraska   Latitude: 40° 49'N
Longitude: 96° 42'W   Altitude: 1166'
Mean Annual Precipitation: 25.7 Inches

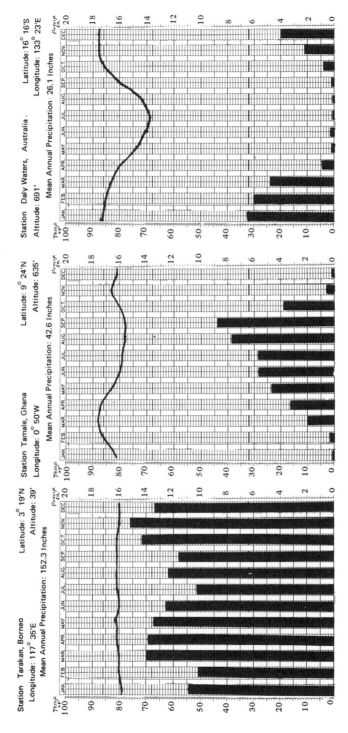

Figure 7.3. Normal temperature and precipitation in eight contrasting climatic situations. The effectiveness of precipitation increases the lower the temperature at which it falls. Very heavy rainfall however, as at Tarakan, Borneo, offsets evaporative moisture loss at high temperatures.

be concentrated in a single wet season. The Congo River rises in its banks as much as 30 feet following the onset of the rainy season.

While many factors combine to affect the extent of moisture loss by evaporation, such as atmospheric moisture content, wind speed, soil moisture, plant cover and global radiation intensity, temperature alone is of great importance. Thus, in the summer season of the middle latitudes when air temperatures are above freezing, the effectiveness of precipitation is primarily reduced by the evaporative power of warm air movement and the water needs of growing plants. Figure 7.3 provides examples of seasonal precipitation in relation to monthly temperatures selected from various climatic regions. In some regions precipitation increases sufficiently during the growing season to offset the water loss by evapotranspiration. In others the summer is dry. The variations are limitless.

The water surplus remaining after depletion by evapo-transpiration is illustrated in Figure 7.4, according to the method introduced by C. W. Thornthwaite.[1] Potential and actual evapotranspiration are shown in re-

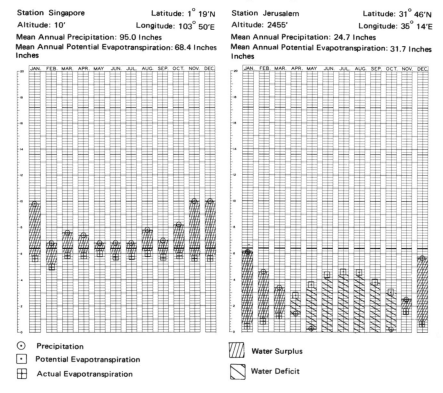

Station Singapore   Latitude: 1° 19′N   Station Jerusalem   Latitude: 31° 46′N

Altitude: 10′   Longitude: 103° 50′E   Altitude: 2455′   Longitude: 35° 14′E

Mean Annual Precipitation: 95.0 Inches   Mean Annual Precipitation: 24.7 Inches

Mean Annual Potential Evapotranspiration: 68.4 Inches   Mean Annual Potential Evapotranspiration: 31.7 Inches

Inches   Inches

⊙   Precipitation

□   Potential Evapotranspiration

⊞   Actual Evapotranspiration

▨   Water Surplus

◇   Water Deficit

1. Thornthwaite, C. W., "An Approach Toward a Rational Classification of Climate" *Geog. Rev.*, vol. 38, 1948, pp. 55-94.

lation to monthly precipitation. Where precipitation is sufficient to supply all the moisture the atmosphere is capable of taking up, as at Singapore, actual and potential evapotranspiration are equal and amounts above the precipitation curve are surplus. When precipitation is insufficient a deficit results, as at Jerusalem.

Vitally affecting the delivery of water for stream transport are the kind, intensity, frequency and duration of precipitating disturbances. In the lower latitudes nearly all precipitation reaching the earth's surface is rain. Most tropical rain falls in short showers of high intensity, whether locally induced by thermal convection or produced by the cumulonimbus towers of a larger, migrating system. Such disturbances occurring over land are often accompanied by high winds of violent gustiness, due in part to the steepening of the lapse rate as land surfaces gain heat during the day. Thunderstorms are exceedingly vigorous and are much more numerous in the low latitudes than elsewhere (Figure 7.5).

In tropical rain forest regions daily temperatures throughout the year average around 80°F and relative humidities range from over 90 percent at dawn to around 70 percent in midafternoon. Copious rain is easily initiated when the atmosphere becomes unstable. The suddenness and intensity of drenching downpours often converts the trails and footpaths on the saturated land to streams and rills of muddy red runoff water in only a few minutes. A representative station in the rain forest of northwestern South America is Andagoya (Istmina), Colombia, in latitude 5° 06′N, about 50 miles inland from the Pacific coast at an altitude of 197 feet. The 15-year rainfall record shows a mean annual amount of 281 inches, falling on an average of 306 days each year. Mean variability is only about 15 percent. Every month averages over 19 inches, with a maximum monthly value of 21.1 inches in April and a minimum of 19.5 inches in both March and December. Rain normally occurs on 21 or more days every month of the year. Each month has recorded more than 5 inches in a 24-hour period. At other tropical stations great rainfall extremes are on record. At Baguio, Philippines, nearly 46 inches of rain have fallen in 24 hours, an amount equal to the entire normal yearly precipitation of many sections of eastern North America. At Plumb Point, Jamaica, 7.8 inches have been recorded in 15 minutes.

In the mid-latitudes thunderstorms and other sudden, heavy showers of short duration are almost entirely confined to the summer season.

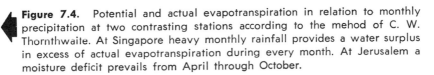

Figure 7.4. Potential and actual evapotranspiration in relation to monthly precipitation at two contrasting stations according to the mehod of C. W. Thornthwaite. At Singapore heavy monthly rainfall provides a water surplus in excess of actual evapotranspiration during every month. At Jerusalem a moisture deficit prevails from April through October.

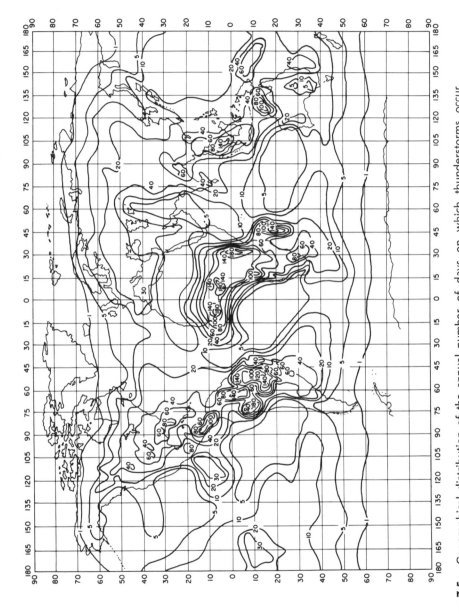

Figure 7.5. Geographical distribution of the annual number of days on which thunderstorms occur. (From U. S. Air Force, **Handbook of Geophysics**, New York, 1961, p. 9.30).

In many areas they become the chief rain-producing events of summertime. But most rainfall outside the tropics is a product of large, migrating frontal disturbances. These are mainly cyclonic systems enduring for 12 hours or more on the average, producing continuous precipitation during the interval of their passing. Intensities typically fluctuate as the spiralling convergence lines of precipitating clouds pass overhead. The intensity, duration and frequency vary widely from region to region, ranging from short, heavy rains to prolonged drizzle. In general it may be stated that most mid-latitude climates experience a summer precipitation increase, with the exception of elongated stretches of continental west coasts poleward of latitude 30 degrees. Here maximum precipitation occurs in winter. Summer rains of east coastal sections in North America and eastern Asia are often greatly augmented by the prolonged, heavy rains of intensely developed tropical storms (Figure 7.6). The *typhoon* of the western Pacific and the *hurricane* of the western Atlantic may occur during any month from late spring into early winter, but usually occur most frequently in late summer.

With increasing distance poleward in the mid-latitudes winters become longer as the period of sub-freezing temperatures lengthens. When rain falls in mid-winter or early spring it frequently accumulates as

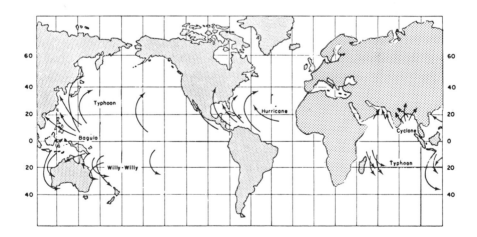

Figure 7.6. Normal tropical storm tracks. Destructive tropical storms are largest and most frequent in the western Pacific north of the equator (typhoons). By contrast, the eastern Pacific and the Atlantic south of the equator do not experience storms of such intensity. (After U. S. Navy Hydrographic Office, **The American Practical Navigator,** Washington, 1965, p. 819).

glaze on impact with frozen solid surfaces. Sometimes known as an *ice storm*, it is more accurately termed *freezing rain*. In the higher middle latitudes and the polar regions beyond, the deep cold of winter frequently supports the occurrence of *ice* showers, a fall of minute, needle-like ice crystals. Frozen precipitation in some form accounts for an ever larger proportion of the year's mean annual amounts with increasing latitude.

The importance of snowfall and other solid precipitation to mid-latitude river systems cannot be overemphasized. About 12 inches of snow, on the average, contains the water equivalent of one inch of rain. The ratio may be 5 to 1 or as much as 50 to 1, depending on the quality of the snow. Accumulations of snow and ice on the frozen ground during winter are released as flowing water in the warm season that follows. Water held firmly on the ground in a solid state is temporarily withheld from the exchange processes until melting begins. The process of water transport is accordingly retarded in winter and augmented when runoff during the spring thaw begins. The work of stream transport in some desert regions and during rainless intervals in many other regions, is supported primarily by the meltwater of alpine ice and snowfields in distant highlands. Heaviest seasonal snowfall accumulations are found in mountainous regions of the middle latitudes. Most of the winter's snow disappears in the succeeding warm season as meltwater for stream flow. But within certain altitudes, year round alpine glaciers and snowfields persist. This is the case for a number of peaks in the Cascades and the Rockies in North America, the Swiss Alps and the Caucasus, as well as the lofty ranges of central Asia and the southern Andes. Persistent snow and ice are found near the equator at altitudes close to 20,000 feet, in the mid-latitudes around 12,000 to 15,000 feet, and at sea level beyond latitudes of around 65 degrees.

Normal yearly snowfall in the United States varies widely between the heavy snows of the western mountains and the much smaller amounts for the rest of the country. Some localities in the Cascades of central Washington and the Rockies in southern Colorado record an average of over 400 inches of snowfall per year, while the coastal plain along the Gulf of Mexico averages less than one inch. In the Cascades of Washington between 400 and 600 inches are expected each winter at elevations from 4,000 to 5,500 feet; in Oregon amounts average between 300 and 500 inches between altitudes of 4,500 and 6,000 feet; in the central Sierra of California about 450 inches are normal at elevations of 6,000 to 8,000 feet. Exceptional amounts have fallen in a single winter. At Tamarack, California, (el. 8,000 feet), 884 inches (over 73 feet) fell during the winter of 1906-07. On the slopes of Mt. Rainier (el. 5,500 feet), the winter of 1955-56 yielded a total fall of 1,000 inches. A single storm may release an enormous fall to a particular mountain locality. For example, at Silver Lake, Colorado (el. 10,200 feet), about 35 miles northwest of Denver, 87 inches of new snow (over 7 feet) fell in a period of 27½ hours in April, 1921.

WATER MOVEMENT ON LAND

The movement of water on land proceeds in a variety of ways that contribute to the vital function of stream systems in the transport of energy, matter and momentum from land to sea. As rain begins to fall, it may collect in ephemeral pools that overflow and drain away in rivulets toward the brooks and streams of an established drainage system. Shallow pools on level land may simply seep into the soil, adding to the subsurface ground water those portions not evaporated back into the atmosphere. The same area, upon drying between rains, may regain a certain amount of subsurface water through the upward movement of capillary action toward the parched surface. On steeply sloping land nearly all the rain reaching the surface may flow rapidly downward into an existing stream course below.

But steepness of slope is only one of several conditions that determine the disposition of precipitation reaching the surface. Also important are general surface configuration, nature of the surface material, whether solid bedrock or unconsolidated regolith, texture and porosity, moisture content prior to precipitation, soil temperature, kind and density of plant cover, and many other less obvious qualities, plus the highly variable characteristics of the precipitation itself. The interrelationships among these conditioning qualities appear in infinite combinations that continually change.

The work of stream transport begins with the impact of rain drops on the surface. Rain drops exert little perceptible effect upon solid bedrock. But where weathering has produced unconsolidated regolith each drop dislodges and disperses a quantity of fine particles upon striking the earth. Once displaced, fine material becomes transportable by tiny rills of flowing water that form as precipitation continues. On sloping land heavy rains often produce sheets of countless interconnecting rills that erode substantial quantities of surface material in a process of *sheet erosion*. Soluble matter is partly taken up in solution. If surface runoff proceeds for a long enough interval, both these products of erosion may continue downward into the flowing waters of an existing stream.

Also contributing to the burden of stream flow is the seepage of ground water. In many areas the continued flow of rivers is possible only through the supply of ground water from an adjoining area at higher elevation. An example of this is the subsurface water supply of the North Canadian River in central Oklahoma. Meltwater from mountain snowfields and glaciers is also an important source of water, suspended particles and dissolved materials for icy streams originating in high mountain regions. This is the case where the Alps supply the upper Rhine, the Rhone, the Danube and the Po, mentioned in Chapter 1. Another minor source of water and stream load is the output of mineral springs.

Streams transport an enormous variety of materials, including countless living organisms, dislodged tree trunks, branches, foliage and other plant debris. But materials derived from rock provide the main burden of matter transported from land to sea. Geologists place them in three main categories: *bed load, dissolved load* and *suspended load*. Bed load is the heavier material rolled and bounced along the stream bed, and it is thought to account for only a fraction of the total load. The particulate matter conveyed into the sea as the suspended load provides the bulk of the burden transported, between 70 percent and 80 percent. For the world as a whole, this is estimated to equal up to 20 billion tons per year. It consists chiefly of the very fine material eroded from exposed soils by direct runoff due to precipitation and is composed largely of clay minerals, silicates and oxides. It is mainly removed from regions of relatively low annual precipitation and is deposited for the most part over the continental shelves and slopes forming the submarine margins of the continents. Maximum yields of sediment apparently originate in areas normally receiving between 10 and 14 inches of annual precipitation, decreasing both above and below these values. Areas with more than 14 inches are generally better protected against runoff erosion by heavier vegetation; areas with less than 10 inches have less runoff.

Dissolved materials are derived chiefly from regions of abundant precipitation and amount to perhaps 4 billion tons annually for the entire globe. The principal components include sodium, potassium, silica and calcium, along with traces of many other elements. Solutes from the land are believed to provide virtually all the ingredients of oceanic salinity. The present mean oceanic salinity of $35°/_{oo}$ is thought to have been reached about 500 million years ago, and is considered to be relatively stable. The capacity of many marine organisms for extracting dissolved substances from sea water and concentrating them to form their skeletons is believed to account for the apparent equilibrium of sea water salinity. Many microscopic marine plants and diatoms, for example, concentrate silica in their minute skeletons. A great many organisms of varying size concentrate calcium carbonate in their skeletons.

The proportion of suspended to dissolved load varies according to the composition of rock material and climate. It also varies greatly from season to season. Table 7.2 presents the proportion of dissolved to sus-

TABLE 7.2

Proportion of Dissolved to Suspended Loads
for Selected Rivers in the United States
at Specific Gaging Stations

(Percent)

River and Station	Dissolved Load	Suspended Load
Juniata R., near Newport, Pa.	66	34
Delaware R., Trenton, N. J.	44	56
Sacramento R., Sacramento, Cal.	44	56
Iowa R., Iowa City, Ia.	30	70
Mississippi R., Red River Landing, La.	27	73
Flint R., near Montezuma, Ga.	25	75
Green R., Green River, Utah	23	77
Mississippi R., north of Alton, Ill.	23	77
Lower Missouri R., near St. Charles, Mo.	15	85
Bighorn R., Kane, Wyo.	12	88
Colorado R., (Texas), near San Saba, Tex.	8	92
Canadian R., near Amarillo, Tex.	3	97
Little Colorado R., Woodruff, Ariz.	2	98

From U. S. G. S. Water-Supply Papers, Quality of Water

pended load for selected streams in the United States. Regional contrasts are illustrated by comparing a river of the arid southwest with one from the humid, forested east. The Little Colorado near Woodruff, Arizona, contains a load consisting almost entirely of suspended particles. The Juniata near Newport, Pennsylvania on the other hand, carries almost twice as much dissolved material as suspended material.

While stream flow is the chief means of water transport from land to sea, a minor proportion moves by underground seepage directly into the coastal waters. This is sometimes accomplished when water-bearing rock strata, *aquifers,* dip downward to produce submarine springs of fresh water. Examples of this are found along the shores of Lebanon, notably near the ancient Phoenician town of Arvad.

Where alpine glaciers reach sea level, as in places along the fiord coasts of the Alaska panhandle and British Columbia, southern Chile and southern New Zealand, water and mineral material are delivered directly to the sea as calving icebergs break off from the main mass. By far the greatest number of icebergs entering the world's oceans in the process of water transport from land to sea are produced from glaciers reaching sea level along the west coast of Greenland. (Tabular icebergs of Antarctica develop from shelf ice at sea level.) Compared with the normal speed of stream flow, coastal glaciers move with ponderous slowness. Still, many glaciers in Greenland move, as they project their tongues into the sea, at a rate of more than 10 meters per day and some reach up to 20 meters per day. Over 70 percent of the icebergs floating southward into the north Atlantic are calved from glaciers along Greenland's west coast between latitudes 69° and 72°N. As they dissolve in the sea, rock materials they contain are released for eventual disposition on the ocean bottom. By this means a number of relatively large boulders have been deposited on the north Atlantic abyssal plains several hundred miles from land.

STREAM TRANSPORT AND CONTINENTAL DRAINAGE

Water transport in all its forms carries energy, matter and momentum from land to sea, and in the process continually alters the setting in which it performs its work. Thus, the effectiveness of water transport changes dynamically at rates that are as diverse as the number of transport systems. Stream flow is only one of the processes in the larger interaction of moisture between land and sea, but it is by all odds the main one.

The relation of stream flow to surface relief features, to the landforms of every land area, is of primary significance in the geography of stream transport from land to sea. In regions of strong relief precipitation increases with altitude. This is especially evident on the flanks of prominent peaks and ranges exposed to frequent disturbances brought against them by prevailing airstreams. Wet, windward slopes commonly stand out in sharp contrast to drier slopes in the lee, creating the rain shadow effect discussed in Chapter 6.

Maximum amounts are mainly concentrated between elevations of 4,000 and 8,000 feet, due in part to the presence of over half the atmosphere's moisture below 8,000. Where the summit height of mountains is less than 8,000 feet, precipitation usually increases upward from base level to a maximum at the peak. This is the case for Mt. Washington, New Hampshire (el. 6,290'), where a yearly precipitation of 70.2 inches compares with 59 inches at Pinkham Notch (el. 2,000') only three miles east, and 36 inches at Bethlehem (el. 1,550'), about 20 miles west. On the tropical island of Kauai in the Hawaiian group, Mt. Waialeale (el. 5,113') at a latitude of about 22°, the average rainfall is 451.1 inches compared with Kalihiwai Reservoir (el. 400') about 10 miles north-northeast with 100.4 inches, and Pali Trail (el. 850'), about 12 miles southwest of the summit where 16 inches is normal.

In many other parts of the world mountainous terrain produces a high concentration of effective precipitation for the support of stream flow. Coastal mountains of northwestern North America, southern Chile, Norway, eastern Malagasy, southern New Zealand, and southwestern India are examples. The striking orographic effect of the Khasi Hills in

Assam, northeast India, upon the northward flux of warm, moist air during the southwest monsoon of summertime is illustrated by comparing Silchar (el. 96') with Cherrapunji (el. 4,309'). Silchar is about 175 miles from the Bay of Bengal and averages 125.5 inches of annual rainfall, 95 percent of which falls in the 8-month period from April through November. Cherrapunji, 75 miles northwest of Silchar, averages 425.1 inches. One of the wetter points on the globe, Cherrapunji has recorded 366 inches in one month (July, 1961) and 1,042 inches in one year, from August, 1960, to July, 1961. However, annual amounts are quite variable, and in 1873 only 283 inches fell.

The headwaters of nearly all the world's major river systems are in mountain country. In some cases a major river receives virtually all its precipitation in the source region of its tributaries, supporting a continuous flow for hundreds of miles through desert where little if any precipitation is added. Such rivers are called *exotic*. The Nile, world's longest river (4,132 miles), is the best example, although the Indus (1,980 miles) in western Pakistan, the Orange River (1,155 miles) in southern Africa, and the Colorado (1,450 miles) are others.

In many arid regions scattered mountain ranges concentrate sufficient precipitation to support the flow of rivers for some distance into the desert basins around them, but insufficient to reach the sea. The Humboldt River in the basin and range country of Nevada is an example. Such regions of internal drainage are subjected to the levelling processes of stream transport but their river systems do not contribute directly to the function of water transport from land to sea.

Estimated volumes of water transported from land to sea over the entire world are scarcely justified for want of adequate data. However, an approximate idea of the magnitudes involved is offered in Table 7.3.

TABLE 7.3

Stream Discharge into the Sea
from Continental Areas

	Water Discharge (cfs)	Dissolved Solids (ppm)	Equivalent Tonnage (ml²/yr)
Asia	12,431,000	142	83
South America	8,962,000	69	73
Africa	6,004,000	121	63
North America	5,100,000	142	92
Europe	2,796,000	182	110
Australia	354,000	59	6
World Total	35,647,000	119 (Mean)	71 (Mean)

From Livingstone, D. A., "Data of Geochemistry," *Chemical Composition of Rivers and Lakes,* Ch. G., U. S. G. S. Prof. Paper 440-G Washington, 1963, pp. 37-40.

Gross quantities for the continental areas and the world total are presented for: (1) water delivered to the sea in cubic feet per second (cfs); (2) dissolved chemical burden in parts per million (ppm); and (3) the equivalent discharge of dissolved substances removed from land in tons per square mile per year. Water moving at the rate of one cubic foot per second is calculated to equal 13.6 inches of rainfall per square mile per year (See Appendix). The total land surface yield of water to the sea per year, according to the table, is 35,647,000 cfs. Asia, the largest landmass, provides, as expected, the maximum of more than 12,000,000 cfs and Australia the least at 354,000 cfs. These are the overall planetary values of the work performed by stream transport from land to sea, minus suspended load, for which estimates are not available.

In Asia about 60 percent of the total runoff volume enters the Indian Ocean and the South China Sea from tropical rivers which take their rise in the southern mountains. Of these the Ganges-Brahmaputra system is the most important, deriving its supply from the steep, well-watered slopes of the eastern Himalayas, where the erosion rate is an estimated 40″/1000 years. Rivers of the Arctic basin, especially the Ob, the Yenesei and the Lena, contribute nearly 20 percent. East Asiatic rivers, particularly the Amur, the Hwang Ho and the Yangtze Kiang, deliver more than 18 percent into the East China Sea.

In South America the Amazon accounts for more than 40 percent of the runoff volume for the entire continent, draining an area of 2,231,000 square miles. Its waters drain into the southern Atlantic, after rising high in the Andes far to the west, its tributaries in some cases originating within less than 100 miles of the Pacific coast. The Orinoco in the north and the Parana in southern South America, each contribute about 6 percent of the total.

Of African rivers the Congo, the main producer of runoff (24 percent), is amply supplied by water surplus from the humid forests and woodlands of Equatorial Africa. The Niger, entering the Gulf of Guinea from western Africa, accounts for less than 5 percent and the Nile only about 1.5 percent.

In North America, the Mississippi system, draining an area of 1,250,000 square miles, produces more than 12 percent of the continental output of water from land to sea. The St. Lawrence, most constant drainage region in North America, produces less than 10 percent, the Columbia less than 7 percent, and the Mackenzie, chief river system entering the Arctic Ocean, provides 5 percent.

In Europe the largest drainage basin is that of the Danube, about 315,000 square miles, and its waters, flowing eastward into the Black Sea, account for 8 percent of the total runoff. The Rhine and the Rhone each produce less than 3 percent. The runoff resulting from Australia's meager rainfall is largely accomplished by a great many small, perennial streams flowing into the Pacific from the eastern highlands. These, plus a small number of northern streams, account for over 90 percent of the output to the sea. Intermittent streams account for the remainder, of which the Murray-Darling system in the southeastern interior is the largest.

Most of the world's continental drainage into the sea enters the combined basins of the Arctic and Atlantic Oceans. By the integrated circulations of these major divisions of the global oceanic system most continental runoff, including suspended and dissolved material, is distributed eventually into the Indian and Pacific basins. This is chiefly accomplished by the exchange mechanisms centering on the circumpolar Antarctic system of currents. It is thus apparent that the present input of elements combining to create the salinity of the sea is distributed in this manner. That the combined volume of the Arctic-Atlantic is very much less than that of the Pacific-Indian Oceans is partly responsible for the slightly higher salinities observed in those waters (see Figure 4.5).

REFERENCES

DAVIS, S. N., and DEWIEST, R. J. M., *Hydrogeology*, New York: John Wiley and Sons, Inc., 1966.

KING, L. C., *The Morphology of the Earth*, 2nd ed., New York: Hafner Publishing Co., 1967.

LEOPOLD, LUNA B., WOLMAN, M. G., and MILLER, J. P., *Fluvial Processes in Geomorphology*, San Francisco: W. H. Freeman and Co., 1964.

RITTER, D. F., "Rates of Denudation", *Jour. Geol. Educ.*, Vol. XV, No. 4, 1967.

U.S. Geological Survey, *Professional Papers*, Many water supply papers, Washington, D.C.

Climate, Vegetation and Soils in the Geosystem

TOPICS

Relation of climate, vegetation
and soils to the geosystem
Geography of climates

Geography of vegetation types
Geography of soils

The spherical earth spins along a helical path through space; one planet in the revolving solar system of planets, satellites, asteroids, comets and meteors. The solar system proceeds on its cosmic course within the Milky Way Galaxy, centrally controlled by the gravitational force of the sun. By its motion in the solar system—diurnal rotation on its axis, annual orbit of the sun—the earth creates the fundamental rhythms with which energy, matter and momentum are exchanged in the geosystem. The geosystem functions in response to the manner in which the planet disposes of radiant energy from the sun. The chief agent of disposition is the atmosphere. The primary process of energy transfer and distribution is initiated at the earth's surface, and the air/sea interface is the main theatre of energy exchange.

The geosystem is the organized interaction of the planet's solid, liquid and gaseous substances within the framework of their present geographical distribution. The details of the geosystem discussed in the preceding chapters concern the present phase in the integration of land, sea and air. The present phase is simply the latest of a very long succession of geosystematic stages normally viewed as the earth's great geological eras, periods and epochs. Dynamically self-modified, each in time yields to a new stage. The present geosystem incorporates an infinite number of subsystems, the vertical dimensions of which are extremely small compared with the extraordinary horizontal reach of their effects.

One-half the earth's sphere is constantly bathed in the sun's rays. By rotating once every 24 hours, one-half the earth is alternately exposed to solar radiation and concealed from it. Diurnal heat gain and heat loss is one of the basic rhythms in the operations of the geosystem. Tangible evidence of the exchange processes operating on a diurnal program is provided in many ways. Among them are the alternate onshore sea breeze of daylight hours and offshore breeze at night; the nocturnal condensation of atmospheric moisture to form dew which is evaporated back into the atmosphere by the warmth and circulation of the succeeding day; the fragmentation of exposed bedrock surfaces at high altitude by alternate heat expansion after sunrise and contraction by cooling at night; and the daily rhythms of temperature increase and decrease that occur within widely different amplitudes on land and water surfaces and in the atmosphere.

Fluctuations of longer period result from the earth's yearly revolution around the sun. Seasonal temperature variations on land, at sea and in the atmosphere are among them. To these may be added the seasonal increase and decrease of precipitation, occurring most conspicuously in the annual airstream reversal of the monsoons; the alternate equatorward advance and poleward retreat of ice and snow; and the yearly rhythm expressed in the middle latitudes by the alternating frost-bound and frost-free state of the soil.

Methodical fluctuations are also provided in the operations of the geosystem by the oceanic tides. The periodic shift of the wave zone of interaction between land and sea and the reversing currents produced by the alternate flood and ebb of the tide are precisions of movement resulting chiefly from the gravitational attraction of the moon.

The fundamental rhythms in the geosystem are profoundly modified by the enormous complexity of the earth's geography. The extent to which the influence of geography is imposed upon the mutual interaction of land, sea and air is definable only within approximate limits.

The complexity of the world's geography has been displayed to some extent in earlier chapters on maps of landforms, of continental drainage systems and of ocean currents. It is also emphasized in the distribution of the world's climates, vegetation and soils.

The climatic diversity of the world (Figure 8.1) reaches a maximum in the northern hemisphere where 67 percent of the earth's land surfaces are exposed to the atmosphere. Large landmasses also extend much farther poleward in the northern latitudes. Thus extreme thermal fluctuations are much greater and more characteristic. So also are extremes of moisture conditions, producing drought-ridden desert on the one hand and humid, rain-drenched forest on the other. The subpolar tundra and taiga climates that span the great breadth of both northern North America and northern Eurasia are lacking south of the equator. A vast arid realm is almost continuous for over 8,000 miles from the Atlantic shores of northern Africa to the Gobi Desert of Mongolia. South of the equator

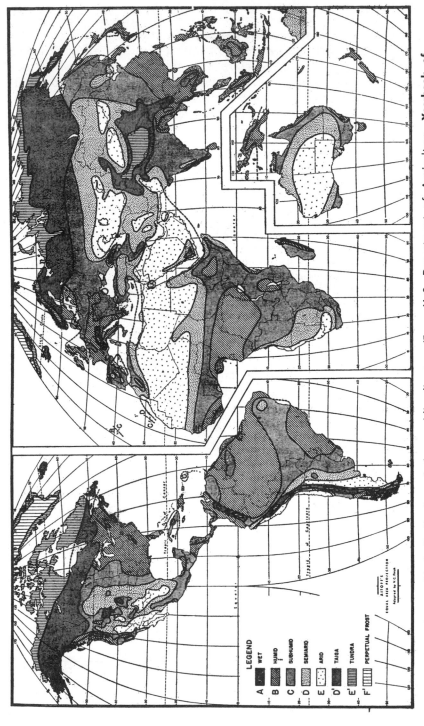

Figure 8.1. Geographical distribution of the world's climates. (From U. S. Department of Agriculture, **Yearbook of Agriculture 1941,** Climate and Man, Washington, 1941, p. 104).

LEGEND

A — WET
B — HUMID
C — SUBHUMID
D — SEMIARID
E — ARID
D' — TAIGA
E' — TUNDRA
F' — PERPETUAL FROST

arid regions are smaller and are separated by large expanses of sea. The largest arid region is in Australia.

A similar diversity of distribution is evident in the world maps of vegetation (Figure 8.2), and soils (Figure 8.3). Plant life and soil distribution are mainly determined by climate. Thus, the patterns presented in the maps of these organic forms are closely analogous to the global array of climates. Distributional diversity is apparent in all three.

The *ice cap* climate of continuous cold takes the tangible form of persistent ice floes over most of the Arctic Ocean nearly at sea level, and over some 85 percent of Greenland's elevated mass. Around the South Pole the enormous plateau of Antarctica, 5,360,000 square miles, is permanently ice-covered, and in addition shelf ice expands greatly during the winter season of June, July and August. Temperatures in the ice cap regions average well below freezing throughout the year, rising above the freezing level only briefly; precipitation is relatively light, increasing toward the equatorward margins; and it is a region without soil.

Fringing the northern reaches of North America and Eurasia the *tundra* shares the deep cold, ice and snow of the ice cap climate in winter, but in summer temperatures rise above freezing for three or four months. Mean temperature of the warmest month remains below 50°F, however. The threat of frost continues from spring until fall, and below the surface the ground remains constantly frozen the year round, a condition called *permafrost*. Precipitation increases in the warmer season as snow and ice melt from the land to reveal a vegetation virtually without trees. Here the many subtle landscape variations are dominated by low shrubs, mosses, lichens, and flowering herbs. Soils are poorly drained, damp and dark with partly decomposed humus.

The boreal forest of taiga climate, largest and northernmost of the world's major forest regions, extends almost entirely across each of the two northern continents. A warm season in which, for three or four months, the thermometer averages above 50°F but not above 65°F allows the development of heavy forest stands in which a small number of coniferous species are the dominants. Winters are severely cold, the coldest months averaging well below zero, and the mean annual temperature range of more than 100° is higher than in any other region in the world. Soils are podzols; that is, generally acidic, with a dark humus layer above a gray leached zone and below that a reddish level of concentrated iron compounds. Scattered through much of the taiga are many deep deposits of partly decomposed plant fragments called *peat*.

The mid-latitude forest regions experience four well-defined seasons, usually abundant year-round precipitation, increasing in summer, and highly changeful weather throughout the year. Normal yearly pre-

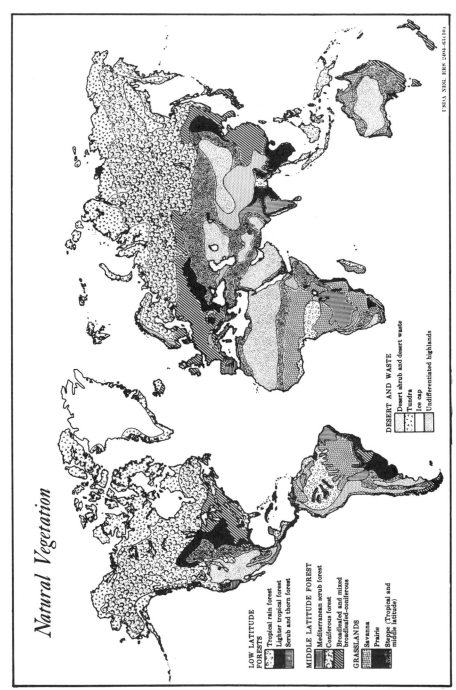

Natural Vegetation

LOW LATITUDE
FORESTS
Tropical rain forest
Lighter tropical forest
Scrub and thorn forest

MIDDLE LATITUDE FOREST
Mediterranean scrub forest
Coniferous forest
Broadleaved and mixed
broadleafed-coniferous

GRASSLANDS
Savanna
Prairie
Steppe (Tropical and
middle latitude)

DESERT AND WASTE
Desert shrub and desert waste
Tundra
Ice cap
Undifferentiated highlands

USDA NEG. ERS 2404-65(10)

Figure 8.2. World distribution of natural vegetation types. (U. S. Department of Agriculture map).

124

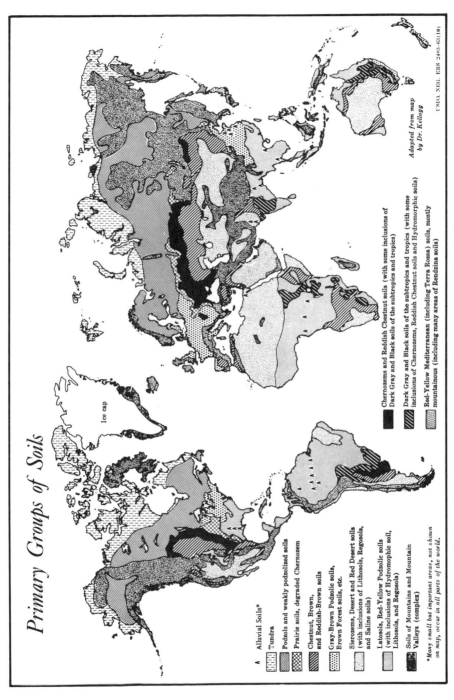

Primary Groups of Soils

▲ Alluvial Soils*

Tundra

Podsols and weakly podzolized soils

Prairie soils, degraded Chernozem

Chestnut, Brown,
and Reddish-Brown soils

Gray-Brown Podzolic soils,
Brown Forest soils, etc.

Sierozems, Desert and Red Desert soils
(with inclusions of Lithosols, Regosols,
and Saline soils)

Latosols, Red-Yellow Podzolic soils
(with inclusions of Hydromorphic soil,
Lithosols, and Regosols)

Soils of Mountains and Mountain
Valleys (complex)

*Many small but important areas, not shown
on map, occur in all parts of the world.

Chernozems and Reddish Chestnut soils (with some inclusions of
Dark Gray and Black soils of the subtropics and tropics)

Dark Gray and Black soils of the subtropics and tropics (with some
inclusions of Chernozems, Reddish Chestnut soils and Hydromorphic soils)

Red-Yellow Mediterranean (including Terra Rossa) soils, mostly
mountainous (including many areas of Rendzina soils)

Ice cap

Adapted from map
by Dr. Kellogg

USDA NEG. ERS 2403-63(10)

Figure 8.3. World distribution of primary soil groups. (U. S. Department of Agriculture map).

cipitation averages between 25 and 45 inches. The growing season is from four to six months long, and mean temperatures of the warmest months range from 65° to over 80°F. The length and severity of winter decreases with diminishing latitude, but the coldest month in most cases averages below freezing. Snow and frost are known throughout the region, and a period of inactivity is imposed upon organic life that decreases in length with decreasing latitude. Forest vegetation dominates in which mixed broadleaf and needleleaf forms, evergreen and deciduous, appear on every hand. Soils are podzolic, ranging from gray-brown to reddish-yellow, and are slightly acidic. Along the continental west coasts poleward of 35° latitude, a narrow, coastal evergreen forest develops. This is a milder region, much more humid, where plentiful precipitation normally increases in the winter season. Mean annual precipitation, largely rain, often exceeds 100 inches. Dense, damp undergrowth and unusually tall coniferous trees form the dominant vegetation, underlain by poorly-drained podzolic soils.

The mediterranean scrub woodland environment is one of dry, hot summers and mild, wet winters with a high percentage of sunshine throughout the year. The growing season is usually six months or more in length, and the warmest months average well above 70°F. The year's precipitation, mainly rain, is concentrated in about six months and amounts to from 15 to 35 inches. Frost and snow normally occur briefly in winter, and the coldest month averages somewhat above freezing. Low, broad-crowned deciduous and evergreen broadleaf trees are typically scattered in an open, park-like woodland, and alternate with nearly treeless grassland and shrub vegetation. The most prevalent vegetation is the relatively dense shrubby *maquis* or *chaparral*. Soils are immature, often partly weathered types that tend to be slightly alkaline, alternating with patches of dark grassland soils.

Grassland regions of the middle latitudes are transitional between humid forest and arid desert. From their wetter margins, where they are usually interpenetrated by outlying tracts of forest, precipitation decreases from around 25 inches to 10 inches or less where they merge indistinctly with desert. The transition is usually evident in the change from wooded, tall-grass prairie through treeless short-grass steppe and scattered bunch grass to the high proportion of bare ground in arid regions of nearly continuous drought. Soils also mark the transition, beginning in the more humid, tall-grass sections with dark brown to black prairie soils that are sometimes over six feet in depth, becoming shallower and less fertile toward the drier limits of grassland. Nearly all mid-latitude grassland soils are distinctly calcareous (alkaline). Four pronounced seasons are typical of these regions, and a pronounced summer precipitation increase is also typical throughout most of the grasslands. Frost penetrates deep into the ground toward the poleward limits of grassland where winter's snows are drifted to lay bare much of the surface. In the northern hemisphere where the largest mid-latitude grassland regions are found, blizzards are an unwanted feature of every winter season.

The essential character of the world's deserts is exceedingly diverse. Arid regions are found near the equator in Somalia and northern Peru, and reach poleward well into the mid-latitudes of North America and central Asia. An exact definition is difficult to present. However, persistent drought arises where the atmosphere's capacity to take up moisture far exceeds its capacity to release it. Deserts persist in many areas in spite of occasional rain. Indeed, showers are frequently seen in some desert areas falling in filaments of gray mist but never reaching the ground. Bright, cloudless, sunny skies prevail during most of the year, and the air is commonly filled with a perceptible haze of dust including minute salt particles. Daily and yearly temperatures fluctuate widely in interior deserts, although coastal regions are moderated by proximity to the sea. In subtropical deserts like the Sahara and interior Australia the thermometer commonly rises to more than 100°F on a high percentage of days during the warmer half of the year. Vegetation is typically scattered and most of the surface is bare. Very low drought-resistant shrub forms called *xerophytes* that survive long intervals without rain, predominate. Soils are thin, immature, very low in organic nutrients, and usually possess a high concentration of salts. They range in color from gray to reddish-yellow and are largely infertile.

Tropical forest and woodland climates are concentrated almost entirely between the two Tropics. They are more or less constantly hot, are free of the threat of frost or snow, and are differentiated chiefly according to the amount and seasonal distribution of rain. *Tropical evergreen rain forest* develops typically where mean monthly temperatures are around 80°F throughout the year and annual precipitation averages about 80 inches (from about 65 to more than 300 inches). High temperatures and humidity prevail. Tall forest trees, often exceeding 150 feet in height are the dominating plant forms. They are mainly broadleaf evergreen. Where a seasonal let-up in the year's rainfall occurs, and with it a seasonal variation in mean monthly temperatures, a somewhat lighter forest of semideciduous trees appears. Intertwined among their branches a higher percentage of trailing vines and lianas is also characteristic. Here mean yearly rainfall is usually lower, but the main difference between this *semideciduous forest* and evergreen rain forest is the partial cessation of rain for an interval each year. Even less rain and a distinct dry season of from two to four months gives rise to a *savanna woodland*. Here the vegetation consists mainly of scattered deciduous trees over dominantly open terrain. Tall, rank tropical grasses are wide-spread, along with a variable proportion of shrubs. Mean monthly temperatures vary more widely than in semideciduous forest, and as the sun rises higher in the sky just at the start of the rainy season, mean monthly temperatures may exceed 90°F. Average yearly rainfall is commonly around 40 inches. Savanna woodland gives way to a *thorn scrub woodland* of fewer, smaller trees, largely deciduous, many of which are armed with rough spines and thorns. Shrubby undergrowth is common and grasses are poorly developed. These are the most arid of the tropical forest and woodland regions, and appear where the annual dry season ranges from four to six months or more. Mean annual rainfall is usually less than 40 inches, and monthly temperatures vary within somewhat wider limits than those of the savanna woodlands.

Soils of the more humid regions in the tropics are called *latosols*. Deprived of a dormant period that annual frost imposes upon other soils outside the tropics, they are continually subjected to the forces of decomposition generated by constant heat and moisture. Moisture deficiency during dry weather in savanna and thorn scrub woodlands varies these circumstances. Latosols are typically reddish due to the high concentrations of iron oxides and compounds of aluminum and manganese. Very little organic humus material accumulates because bacterial decomposition is seldom at rest.

Highland regions of the world possess a variety of environments that create a pattern of enormous complexity. This is particularly true of the lower latitudes where beginning in the adjacent lowlands a change upward from tropical evergreen rain forest to permanent ice and snow may be encountered within a vertical distance of about 20,000 feet. An unusual environment found on strong relief features in the tropics is the *cloud forest,* or *mist forest,* sometimes known as a *moss forest.* This develops where clouds frequently bank up against a mountain side and thus drench the sloping surface with mist and rain during much of the year. These dense, evergreen vegetation forms are commonly found between 5,000 and 8,000 feet.

Climate, vegetation and soils are produced by the continuous processes of the geosystem. These processes operate primarily according to the periodic influence of the earth's motions. The rhythmic functions of the geosystem are reflected to some extent in the pattern of climates, dominant plant associations and soils, in which a degree of symmetry is seen in the maps, Figures 8.1 - 8.3. The low-latitude regions of persistent heat and humidity that are dominated by tropical forest and woodland and are underlain by reddish, lateritic soils, are almost entirely confined within 25 degrees of the equator. The poleward transition from these terminal climates, especially in the western hemisphere, Af-

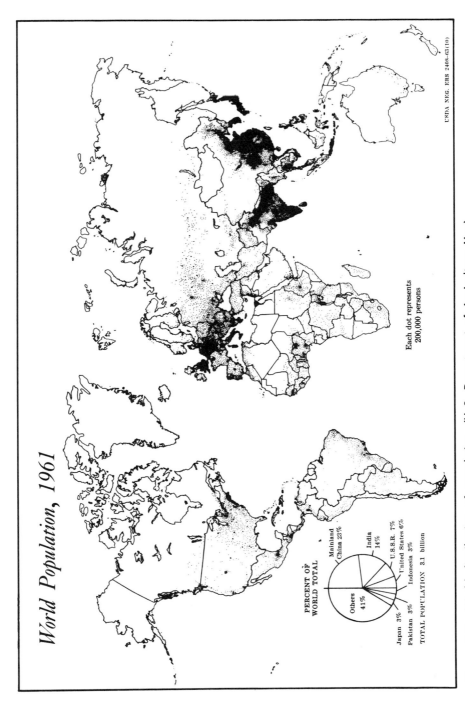

Figure 8.4. World distribution of population. (U. S. Department of Agriculture Map).

World Population, 1961

Each dot represents
200,000 persons

PERCENT OF
WORLD TOTAL

Mainland
China 23%
India
14%
U.S.S.R. 7%
United States 6%
Indonesia 3%
Japan 3%
Pakistan 3%
Others
41%

TOTAL POPULATION 3.1 billion

USDA NEG. ERS 2408-63(10)

rica and Australia, is toward subtropical desert which is mainly confined between 25 and 35 degrees of latitude on both sides of the equator. Mediterranean regions distinguished by dry, hot summers and mild, wet winters, are found in widely separated segments on the west coasts of the major landmasses approximately between 30 and 35 degrees latitude. Equatorward from each of them is coastal desert, varying in length, but in all cases washed by the equatorward flow of relatively cold oceanic currents (see Figure 4.1). Although contrasting greatly in size, shape and intrinsic character, these regions of climate, vegetation and soils are so arranged as to indicate the partial attainment of a balanced distribution of the processes operating in the geosystem.

But the distribution of man is very much more uneven (Figure 8.4). And his connection with the environments created by the geosystem ranges from the consistently successful to the chronically disastrous. Examples of the former are the world's great grain and livestock regions such as the U.S. Corn Belt, the Russian Ukraine and the Argentine Pampa. Examples of the latter are the vastly over-populated regions of monsoon Asia like the North China Plain, the Kwanto Plain of Japan and the Bengal Delta of India and East Pakistan.

Where man has ignored the limits within which the environment is capable of supporting him continuously he has made tenuous his own survival. But where he has carefully assessed nature's capacity to accommodate him he faces a future of indefinite prosperity.

The geosystem, in which change is inherent, offers man environmental conditions of unending variation. But variation within foreseeable limits reckoned in terms of a human life span. Man, on the other hand, has reached a stage of technological progress in which his ability to alter his environment has exceeded his capacity to keep his activities in harmony with the dictates of nature. The geosystem will encounter no difficulty in overcoming the geographical imbalances imposed by man. But only at its own terms and at incalculable cost to man. Thus the inevitable course man must follow requires an accurate appraisal and understanding of the substances and processes of geosystematic operation, an unqualified willingness to live within the bounds set by those operations, and a realistic plan by which all mankind may flourish within those bounds.

REFERENCES

BERG, L. S., *Natural Regions of the USSR*, New York: The Macmillan Company, 1953.

GLEASON, H. A., and CRONQUIST, A., *The Natural Geography of Plants*, New York: Columbia University Press, 1964.

GOODE, R., *The Geography of the Flowering Plants,* 3rd ed., New York: John Wiley and Sons, Inc., 1964.

HADLOW, L., *Climate, Vegetation and Man,* New York: The Philosophical Library, 1953.

JAMES, P. E., *A Geography of Man,* 3rd ed., Boston: Blaisdell Publishing Co., 1966.

RUMNEY, G. R., *Climatology and the World's Climates,* New York: The Macmillan Co., 1968.

Appendix

CONVERSION TABLES

Temperature

Celsius	Fahrenheit
50°	122°
40	104
30	86
20	68
10	50
0	32
− 10	14
− 20	− 4
− 30	− 22
− 40	− 40
− 50	− 58
− 60	− 76
− 70	− 94

Distance

1 nautical mile	6,076 feet	1,852 meters	1.85 kilometers
1 statute mile	5,280 feet	1,609 meters	1.6 kilometers
1 kilometer	3,281 feet	1,000 meters	.62 statute mile
1 meter	3.28 feet	100 centimeters	39.4 inches
1 centimeter	10 millimeters		
1 fathom	6 feet	1.83 meters	

Speed

1 knot	1 nautical mile per hour	
1 meter per second	3.28 feet per second	2.24 miles per hour
1 foot per second	.31 meters per second	

Precipitation/Runoff

1 cubic foot per second	643,000 gallons per day	13.6 inches rainfall per square mile per year
1,000,000 gallons per day	1.547 ft³/sec	21.002 inches/ml²/yr

Index